CRAZY ME

How I Lost Reality and Found Myself

By D. Thomas Bixby

CRAZY ME
How I Lost Reality and Found Myself

Micro Publishing Media, Inc.
PO Box 1522, Stockbridge, MA 01262
www.myserenitypress.com
info@micropublishingmedia.com

Published by

A Division of Micro Publishing Media, Inc.

ISBN 978-1-944068-58-5

This book is not intended as a substitute for the medical advice of physicians. The reader should regularly consult a physician in matters relating to his/her health and particularly with respect to any symptoms that may require diagnosis or medical attention.

The material in this publication reflects the author's perceptions of his personal experiences as drawn from memory. These perceptions may differ from the way other persons remember the same incidents or from objective facts. Some names and other details have been changed to protect anonymity.

Although the author and publisher have made every effort to ensure that the information in this book was correct at press time, the author and publisher do not assume and hereby disclaim any liability to any party for any loss, damage, or disruption caused by errors or omissions, whether such errors or omissions result from negligence, accident, or any other cause.

Front Cover by Michael Yuen-Killick
Book design by Jane McWhorter

Printed in the United States of America

CONTENTS

PROLOGUE
I Can't Remember Dying...

I clung to the feeling of having a body with all my might but the truth, the awful and inescapable truth was that I was already dead. My body, in fact not only my body but everything I saw or touched or heard, everything in the world was no more than a dream of sorts, an illusion projected into my mind by the great evil that was the source of all creation.

Liquid terror ran through my veins forcing a powerful and pungent acid through every capillary and into every muscle and every single cell of bone, nerve and flesh in my weakened body. The acid was eating away the final remnants of my solid existence. I would soon dissolve and my naked soul would be exposed to the true horrors of eternal damnation.

When I had tried to sleep some hours before, something cold and evil had invaded me. It was dark, not like the darkness you feel when you turn off the lamp or when you cover your eyes with your hands. This was something deeper, more sinister. It was like somebody had turned out my life, turned off the whole universe, but somehow I was still there… alone… disembodied… watching… nothing!

The drapes, the bed, even the walls . . . they seemed unreal . . . like they were made of some kind of black, watery substance. I couldn't tell where I ended and where the walls began. It felt as if, without any warning, some great hand could just stir the water; and the walls, the whole of reality could just ripple and dissolve into something else… anything else… and me, I could dissolve with it… into something else… anything else.

I buried my face in the pillow. I clenched my fists and tensed my legs and ground my teeth together struggling desperately to cling to some feeling of an earthly body. I was afraid to pull back the drapes on the bedroom window or to open the door to the hallway, fearing I would not see the wooded lot behind the house or the upstairs hallway but rather the black and eternal void that awaited me.

It was June of 1970. I was twenty-two and a half years old. I was slipping into the unstable and terrifying world of schizophrenia. Much of my life up to that point had been anything but ordinary. I had had the worst cases of all three childhood diseases the doctors said they had ever seen and my heart had stopped from the drugs they administered to fight off convulsions and brain meningitis from the mumps. When I entered second grade I weighed only twenty-seven pounds compared to the fifty-five to seventy-five that was normal. During my grammar school years, I suffered through four years of brutal and daily beatings at the hands of three street toughs, each of which was a head taller and a good thirty pounds heavier than me. In junior high, sick for years of being bullied, I dared to stand up to a gang of kids who were pushing other students around. As a result, I was stalked and threatened by them for almost two years. At home I was sometimes subjected to whippings with a folded over belt or a wet dishrag.

By the time I left high school I was already a heavy episodic drinker and as a result survived two automobile accidents that totaled both of the cars I had been driving. As a knee jerk reaction to the bullying I had received I joined the Army and became possibly the world's skinniest Military Policeman. During the 1967 riot at the Fort Dix, NJ stockade I was one of the first thirty-five MP's sent in against several hundred prisoners armed with pipes, rocks and potatoes studded with razor blades. Soon after, I was put on standby at the Newark race riots. In the Vietnam War, I served both as an MP and as a volunteer helicopter door gunner, flying more than one hundred missions over enemy territory. At one point I was thrown from the violently bucking ship and left dangling by my harness three thousand feet above the South China Sea. As an MP in Vietnam I saw hundreds killed in the highly publicized 1968 Tet offensive and later lived in a small outpost that was attacked every night for several months running. I patrolled roads that were frequently fired on by snipers and subject to ambushes and mining.

Near the end of my tour I was nearly drowned trying to retrieve the stiff and rotting corpse of a fellow soldier from a muddy river filled with razor wire and explosive mines. I survived two ammunition dump fires, shrapnel wounds in the leg and was hospitalized three times for malaria symptoms and gastroenteritis. During the worst episode I suffered a fever in excess of

106 degrees which is often fatal. There were no helicopters available and I was placed in the back of a truck, delirious and weakened to the point of paralysis, and driven sixty miles over dirt roads to an evacuation hospital. I witnessed multiple deaths by enemy attacks and by accidental friendly fire. I saw a man on a bicycle crushed by a tank, a four-year-old girl killed by a five-ton truck, a six-month old infant killed by a hand grenade and I lived in a bunker frequented by snakes that killed in seven seconds. Every one of the thirty Vietnamese National Policemen, that I had lived with and fought alongside at the tiny outpost, were hunted down and executed by the communists after I came home. Many of their families were murdered along with them.

At age twenty-two I was no stranger to fear or to evil, but never in those sometimes turbulent years had I felt fear, abject terror or the presence of evil like I did on that warm, calm summer night in 1970. Never, in spite of all of the things I had lived through up to that point, had I felt the absolute hopelessness and infinite paranoia that filled me to the depth of my soul throughout that sleepless night in June, and then… it got worse.

People have asked me what it's like to be schizophrenic. I tell them to remember their worst nightmares and then to take a few minutes and think of all the ways they could have been worse. Now I tell them to imagine that every time they think of a way it could be worse, it happens… not in their dreams but in their waking lives… and it goes on for years.

My schizophrenia was that bad. It was an overwhelming and deeply pervasive fusion of terror and despair, of frustration, disorientation and mental anarchy that gnawed and bored into the most hidden areas of my mind, destroying even the most fleeting sense of joy or peace in my body and my soul.

There were many times during the first three years of my schizophrenia that I would most certainly have killed myself except that I sincerely feared I was already dead and living in some realm or other of hell. I was afraid that if I tried to escape the horror by suicide … I would only wake up in a far worse and more hellish nightmare. There were times that I would have traded my disease of the soul for any disease of the body or anything that would offer a certain and final death of my mind.

I did not kill myself and instead somehow found the strength to fight on through years of acute and sometimes chronic symptoms. I became lost for months at a time in a part of my own mind obsessed with unraveling the nature of reality and its seemingly evil origins. My body would sometimes actually convulse with waves of terror and at other times feel dead to me as I would sit and stare, motionless for hours.

Initial drug therapies administered by the first psychiatrist I sought out caused serious side effects which massively worsened my condition. This was one of many setbacks and relapses I suffered as I attempted to claw my way back to some sense of normalcy. I also achieved several breakthroughs along the way and many of these were due to interventions and therapies that are to this day outside the medical models of treatment for the disease.

I did not kill myself as more than ten percent of schizophrenics do nor did I end up permanently institutionalized as was considered at one point. Neither do I have to be maintained on the powerful anti-psychotic drugs that were initially given me and in my view actually prevented the more complete breakdown of my defenses that was ultimately necessary for the healing and rebuilding process to begin.

I live a full life and have for many years. I believe it will come as a great surprise to many that have known me for some time through business or other pursuits that I was, and by medical definition, am still considered schizophrenic.

When I wake now, it is without the horrible gnawing terror or the evil blackness that haunted me for so terribly long. Small joys and moments of peace take on proportions that were unimaginable not only during the worst of the disease, but also went unnoticed or unappreciated before I fell ill.

I can remember the very day and the very moment that I felt a first ray of hope that what I was experiencing was not eternal damnation but an illness from which there was at least a chance of recovering. It would still take many, many months before I would become reasonably convinced that the world was something other than an empty illusion in which every person, place or thing, every song on the radio, every spoken word or sign on the highway was part of a giant conspiratorial projection orchestrated by the great evil that was behind all things, the sole purpose of which was to draw me nearer and nearer to the ultimate and eternal black void of hell that was my inescapable destiny.

I had been given up on by two psychiatrists. The second one passed me off to a staff psychologist who was working in his office. It was during my third visit with her when that first glimmer of hope emerged. The other doctors had merely listened, sometimes impatiently, as I tried to describe the horrors of the last few weeks. Eventually they would interrupt me to offer a change in medication, or to give insight as to how some other person in my condition had submitted to electroconvulsive shock therapy and was more at peace as a result.

The psychologist, Dr. Jane Rittmayer, had listened much as the other two doctors, in my first couple of sessions with her as I went over the fear and other symptoms that were most on my mind. During the third session I confessed my deepest fears, that I was not truly alive and that even she could be part of the illusion or dream I was experiencing. I was looking down at the floor as I talked and mumbling to an extent as if talking to myself as I wasn't completely convinced that anything or anyone else including the doctor was real. When I paused she spoke softly, to some extent tentatively as if she wasn't certain whether she was about to say the right thing.

She told me that she had had a similar experience. She said, "It wasn't exactly the same as yours, but it was similar." She went on to tell me she had come to believe she was a witch and that deaths and other occurrences around her were being caused by her thoughts. I could tell from her tone that her experience had been as real and terrifying as my own and I could tell from her demeanor that she was no longer under the spell of it.

It was that very moment when Dr. Jane shared her own dark secret that started me on my long road to recovery. Once I knew another human being had shared in some fashion the same pain and fear that I was experiencing and that she had somehow moved beyond it, I could dare to hope.

I am hopeful that by communicating my own experience that someone who still suffers may gain hope, as I did. What I will describe is a physical and mental journey, a deeply emotional journey and most of all a spiritual journey which is at first exhausting and hopeless, then ultimately uplifting. I think of my battle with schizophrenia much in the way I think of my experience in the Vietnam War. I would never wish even the smallest part of it on anyone and I hope to God to never go through any of it again, but having gone through it, I am much the better for it. By comparison to what I lived through, I can find small joys in even the simplest, most ordinary situations and I can find a degree of peace in even the most difficult of circumstances.

I will share with you many of the intimate details of the way my mind worked during my disease and also a sense of the emotional tempest that tossed me upside down and turned me inside out. There were physical issues also and many of them were caused by the medication side effects. I will tell you about the events that may have led to my breakdown and also about the roller coaster of exhilaration and despair as I began the long climb back from my personal hell.

I have experienced schizophrenia from the inside out. What I know about it comes mainly from that. Many of my own beliefs about the nature of my disease, its origin and the means for recovery are well outside of what is currently believed in the medical field. When Kurt Vonnegut's book *The Eden Express* was published, many who subscribed to the then current medical model of schizophrenia were critical even to the point of denying that he had even been schizophrenic, simply because he had recovered. Scientific thought on what constitutes schizophrenia changes often. Thought on what even constitutes science has changed radically especially in the century since the introduction of Quantum Physics. Popular theories about causes of schizophrenia have ranged from demon possession to early family trauma to chemical imbalance. Even the most current medical or scientific models of the disease offer little hope for a full recovery to the great majority. I am not hopeful that sharing my experience will change any but the most open of medical minds even though I believe much of what is commonly believed about schizophrenia is dead wrong and that many of the medical interventions are counterproductive.

I am hopeful, however, that someone who suffers as I have or their loved ones may gain some small measure of hope that it is possible to recover and that they will be able to break free of the debilitating fear and despair and live again, as I have.

CHAPTER 1
Machinations

There were, I believe, many reasons for my breakdown and looking back it is hard to deny that the seeds of my "differentness" may have been present very early in my life. A single moment of awareness of a ten-minute gap in my consciousness while driving however; was the beginning of a vicious mental spin cycle that tied me in knots for months at a time and clearly contributed to my fall.

"Who in the hell has been driving the car for the last ten minutes?" I thought to myself. I tightened my grip on the steering wheel of my little forest green 1969 Volkswagen bug. We were nearly to Peddler's Village in Lahaska, Pennsylvania and the last I remembered thinking anything remotely related to driving was back on River Road between Trenton and Lambertville, New Jersey. I had been lost in some extreme reverie, off thinking about some idea or image that had been triggered by a song on the eight track tape player or by something one of the old Ewing High School friends riding along with me had said. The distraction had been so complete that my conscious stream of thought had left the task of driving entirely and it was a serious jolt to realize that someone or something, other than who I knew as myself, had to have been driving the car. I began to question in a profound way, who I really was, what role my consciousness played in perceiving the world and ultimately whether the outside world existed at all.

I was not always a good student in spite of, or perhaps because of the fact that I was always a deep thinker. In 1954 my family returned to Trenton, New Jersey after four years on the west coast, two in the San Francisco area, and two in Seattle. I was shifted mid year from the public school system in

Washington to a Catholic grammar school, Blessed Sacrament, in Trenton. It was a tremendously disorienting move. The Catholic kids were already well into writing in cursive using the Palmer Method while at Coe School in Seattle we had barely just learned to print. We also had to use fountain pens to write with at Blessed Sacrament when I had never used anything but a pencil.

Sister Mary Assumpta had little patience for my inability to keep up with the rest of the second graders who all had a year and a half head start on me in both Palmer method and writing with ink. I had absolutely no ability to achieve the loose grip and flowing stroke which was apparently necessary to create something other than the awkward, broken and disjointed lines that I scrawled out when answering the catechism questions that were fired at us several times each day. My tension worsened when I learned the nuns would sneak up behind students and grab their fountain pens. If a pen did not slip easily from a student's grip, if they had been holding on too tightly, as was invariably my case, the Sisters would whack their knuckles with the flat side of a ruler.

The tension in my writing hand was of course in part due to my fear of not knowing the cursive letters or how to use a pen but it was also an extension of the tension in the rest of my body. Back on the West Coast I was used to wearing corduroy pants with a flannel shirt or khakis and a t-shirt to school, depending on the season. Scratchy wool pants, rigid leather shoes, a starched white shirt, thick tweed suit jacket and a strangulation necktie were all part of the required year round uniform at Blessed Sacrament. We also carried heavy leather briefcases containing six textbooks back and forth to our homes every day to insure that we were studying each and every subject, every night.

The lightest of those books was the Catholic catechism which ironically contained the weightiest of the subjects… but not to Sister Assumpta, and this is quite possibly where my tendency toward deeper thinking first became a problem for me. For Sister, catechism class was extremely simple and straightforward. She asked the questions and the class responded out loud in unison with the answer they had memorized from the mercifully thin and lightweight text they had carried home the night before.

"Who made me?" Sister would ask. "God made me," the class would chime back. When Sister thought someone was lagging behind the program she would call them out individually.

"Who made you, Mr. Bixby?" In Catholic school I was no longer Tom or Tommy as I had been for the first seven years of my life. I was now Mister

Bixby. That would have gone along nicely with the image created by the suit jacket, tie and briefcase, except for the fact that I weighed less than thirty pounds and looked half frightened out of my wits by this stern witch dressed in black who kept threatening us with a yardstick.

"Well, it says here in the book, that God made me, but..." There was an audible gasp from the other students. Sister Assumpta glanced down to find she was holding the catechism instead of her ruler. "Who made you, Mr. Bixby?" She repeated herself loudly as she took a step closer toward me. I took a breath expecting the forces of right and might to appear magically and explain that my dilemma in thinking the subject possibly more complex than the way it was being presented was not unique and that perhaps some additional thought or explanation was warranted.

As I paused momentarily trying to think how best to phrase my confusion, Sister rolled the catechism into a tube as she again took a step closer. My deeper thinking self fled into a yet deeper part of my psyche, awaiting another time and place to sort out right from wrong or how we could possibly know for certain that God made us. "God made me, Sister," I half whispered and half choked. She turned to walk away, then keeping her back to us; she looked up at the crucifix on the wall above the blackboard. "Who made me?" She shouted out. I was sure of it now. "God made me," we all shouted back. Then Sister Mary Assumpta turned back to us and smiled sweetly. It was at about that time in my life that the facial ticks started, as well as the nightly bed wetting.

I had sister Assumpta again two years later in fourth grade. I began to excel in certain subjects like mathematics and when it came to diagramming sentences and identifying even the most complex parts of speech, I was at the top of the class. Still, there were many areas which made less sense to me but the part of me that wanted to question or beg for more explanation had already been silenced by the pressure to conform and by the desire not to be hit more than necessary.

In religion class in fourth grade we were taught that people have souls but that animals do not, nor do animals think. I could buy that our minds were different from animals, but the idea that we had consciousness and animals had none, that animals acted merely on a kind of robotic programmed instinct seemed off to me. My intuitive sense of things screamed "wrong!" When questioned about such things, the nuns would either force us to repeat the tenets until we sounded as if we believed them or they would simply explain that the reason we know such things is "because we have Faith."

We were continually programmed by the threats of physical violence or of punishment in the afterlife to accept what was put before us as fact. The more I was asked to accept things on faith, the more a part of me separated from the socially engaged part of my personality and began to hide out somewhere else.

During the failed 1956 Hungarian Revolution in which rebels attempted to recapture their freedom from occupying Soviet Communist forces, Sister Assumpta stood in front of the class and informed us that the rebels had all been Catholics under the direction of the saintly Cardinal Josef Mindszenty, and that the atheist Communists had gained the upper hand and brought the rebels to a central collecting point where they were being processed through a giant meat grinder! The remains we were told were separated, the bones were floated down the river and the flesh and blood was shipped back to Russia to be used as fertilizer on the collective farms under the Five Year Plan!

Stories like that still triggered some kind of alarm in me. I knew somehow to question what I was being told but I had already learned to keep my questions inside, to be answered at some later time. Sister Asumpta once offered an explanation as to why Puerto Ricans carried knives. Sister apparently felt this was a question she needed to answer even though no one had raised the subject. "You'd carry a knife too, if someone charged you a quarter for an apple." She pinched up her face accusingly as she answered the question she had raised herself.

On another day Sister Assumpta ranted on about a friend of hers who had suffered convulsions after drinking pickle juice. I remember being quite confused by the intensity of her rambling and the disgust and revulsion she displayed about the incident. It was many years later when it finally struck me like a bolt from the heavens that Sister's story may have been a parable of sorts about a young lady who was punished by God for partaking of another item that was shaped similarly to a pickle.

Tyrone Power, the movie idol, we were told, was in hell as he had had a heart attack and fallen off a ladder to his death while filming a dueling scene which naturally would not have allowed him time to make an act of contrition for all the revoltingly adulterous behavior he had taken part in during his sinful life. This insight was offered along with a long list of the various ways we could die or be killed on the way home from school that afternoon in order to motivate us to memorize a prayer called The Act of Contrition. This was a prayer which we needed to have ready at any moment as a defense against the flesh eating fires of hell in the event we were crushed under the wheels of a school bus or knifed by a Puerto Rican who was understandably angry about being charged a quarter for an apple.

Mercifully and finally I was expelled from Blessed Sacrament at the end of the sixth grade term. I had dared to question and in fact refused to complete a ridiculous punishment task I had been assigned. I was convinced that not just the specifics but the entire punishment was unjustified. All I had done was to stand up to one of the many bullies that had humiliated and beaten me mercilessly for years while I attended Catholic school.

Sister Berkman, the school principal, presented me with the dreaded "transfer" to the public school system where she implied I would be more at home with the other communists and children who were under the spell of Lucifer. She told my mother I was the smallest child in the sixth grade but more obstinate than the biggest of her eighth graders. My mother cried. My father, who was a Methodist, was less upset.

Years later he told me quite oppositely that he thought being expelled from Catholic school was the worst thing that ever happened to me. This was another of the conflicting and confusing messages that formed the basis for my somewhat tentative grip on reality. When the incident in the car on the way to Peddler's Village shocked me into questioning how I looked at the world and in fact who or what I really was, there was indeed much to be questioned.

CHAPTER 2
Eternity

When Stephen Hawking published “A Brief History of Time” in 1974, the Big Bang Theory was not a new idea to me. During the breakdown, I thought deeply about the possible origins of the universe and one of the ideas that came to me was an exact replica of the explosive vision Hawking presented four years later. What had started my deep investigation into the nature and origins of both my internal world and the external universe was a straightforward and simple question that kept coming back to me as I made my way through the gaudily overstocked and over decorated gift shops of Peddler’s Village that chilly evening. “Who the hell was driving my car while I, or whoever I previously thought I was, was off somewhere else?”

The first and most obvious answer that came to mind was in line with Freudian notions of the unconscious. While that seemed an easy explanation, it only increased my anxiety. Was there someone else living inside me? Another entire being, that was complete enough to take over my body and manage complex operations like driving? If that was true who was the real “I?” What was my function? Was I what I thought I was, or just along for the ride? Who made my decisions for me? Was all the thinking I did on the surface just overruled by or irrelevant to some deeper mind that resided inside me? Now my anxiety turned to fear as another idea froze me in my tracks. What if that “other” wasn’t entirely inside my brain? What if it was connected to something outside of me, some larger, even cosmic mind? Maybe my mind was connected to other people as well!

That might explain the strong feelings I got when I sensed someone was lying to me. These intuitions often made it hard for me to function simply like most people. When I felt my parents were lying to me or when it seemed like a teacher wasn't telling me the whole story, something in me grabbed my attention and caused me to question and often to confront them. At home when I dared to question my parents' real motivations about something or the way in which they mistakenly remembered something, I was met with icy and steadfast rebukes. If I pushed an issue my father would explode into a brief but blinding rage. "Mary get his pants down. I'll get the belt!"

My mother was coolly controlled and as sweet as syrup when my father or anyone else was around to observe her but it was a different story when my sister and I were home alone with her. She had no patience for anyone that questioned her or upset her obsessively detailed routines around the house. I believe my mother had the same intuitions or deeper sense of things that I had but kept herself completely unaware of it by micromanaging every second and every bit of minutia in her life. If I put the raisin bran back in the cupboard to the left of the corn flakes instead of the other way I was in for a focused and sometimes heated lesson on why her way was the only correct way. When I did question or challenge her perceptions she responded with a quietly threatening tone, sometimes ending the matter before it could deteriorate into an actual discussion with a slap, pinch, an elbow or on rare occasion with a wet dishrag across the face. I ran my forehead into the edge of a sideboard in the dining room trying to escape from her after an argument. I saw stars and fell to the ground clutching my head and my eye. She stopped short in the doorway between the kitchen and dining room and continuing to wipe dry some utensil with a dishtowel she spat out, "See. God punished you for talking back to your mother."

A deeper or different sense of things continued to build in me. I often felt I knew or could somehow see things were not as they were being presented. It contributed to my feeling of being an outsider, but because I got such strong reactions when I raised questions, I also learned to subdue my own honest reactions and to fit into almost any situation or within any group.

Later, as I struggled with the questions that had been raised by the trip to Peddler's Village, I became aware of another conscious entity within me. I realized that the thoughts and images that flew through my mind, the questions, pictures, words, in fact the constant dialogue that was raging was comprised of not only the roles of questioner and the provider of possible answers but yet another entity that watched the entire process unfold but was somehow completely neutral or divorced from it. This was a second big jolt that helped to knock me out of the normal stream of thought I was used to following along never questioning whether "that" was "me." As I began to disassociate with the thoughts that streamed across the projector screen of my mind I became more and more aware of this quieter but far broader sense of self which I named "the watcher."

I began to separate from the thoughts and other content of my mind and became able to watch them as they arose and fell. I was still there or rather had always been, but the thoughts I had unquestioningly believed were me, were now over there, somehow separate from what now seemed a truer me. As my quest for who I really was continued over many weeks I became lost in deeper and deeper thought until I began to lose some of my connection to the external world. I had been released from the Army to attend Trenton State College about a year after I returned from the Vietnam War. I had done well during the first semester, receiving four A's and one B. I had taken basic liberal arts courses in Philosophy, Language and Communications, Math, Meteorology and Sociology.

During the second semester, which began in January of 1970, I began to falter in my classes and started to have episodes where I would become disoriented. Walking across the campus quad I felt completely lost as if I had just been dropped there from another time or another planet. I knew I was on my way to class but I couldn't remember which class. I was in a fog, so much so that I began to question whether I should be following the schedule from the current semester or the one before. Rather than sort through the confusion or seek help I turned and went back to my car and left campus. I began to have frequent dreams about being disoriented on campus, not knowing who I was or where I was headed. My disorientation was so complete that looking back now, I cannot remember which episodes were dreams and which actually happened on campus while I was awake. My confusion only increased my efforts to become certain of who I was at the deepest level. These mental machinations increased my distance from schoolwork and the other things which might have helped to ground me in the outside world.

I doubled my efforts to explain the ten-minute gap in my consciousness while driving. Maybe I had never actually driven the several miles between New Jersey and Pennsylvania. Maybe I had experienced a time warp or a distortion of space. Maybe time and space were not as constant or concrete as they appeared. The Chinese sage Lao Tzu had written that there was no way he could know whether he was a human having a dream of being a butterfly or a butterfly dreaming it was human. What if time and space and all the forms we know as reality were nothing but dreams? That would certainly explain the gap or the moments of disorientation that I had been experiencing.

Maybe I was someplace else altogether, still unconscious in the hospital in Quinhon, half dead from fever. I had felt like I was dying in the river in Bong Son. Maybe I had died. Maybe my body had long ago caught up in the razor wire by the bridge and been blown to bits by the mines. Maybe my mind had escaped at the last minute into a kind of dream world and everything I had experienced since was no more than a desperate attempt to deny the reality of my death.

What was real and as importantly, what was the instrument, or part of me I used to perceive and evaluate the world? I read about movies. They are called motion pictures but the truth is that in all the movies we have ever seen we have never once seen any actual movement... none. We sit in the dark as a series of still pictures are projected onto a screen in front of us. The movement we perceive is all assumed by our brain! In fact, during a two-hour movie we actually see pictures for only about an hour. During the other half of the time we are in darkness while the projector shutter is closed as the mechanism pulls the next still picture into place.

I realized that even in the most directly experienced situation there was actually a time delay between the event in real time and our perception of it. Light and other stimuli coming from an object take some time to travel to our sensory organs. There is a further delay as impulses travel along the nervous system to various receptors in the brain. The time delay for light to reach us from the sun is eight minutes. Sound takes about a quarter of a second to travel the length of a football field. Even though the time lag from event to perception is very short in most cases, it was still troubling to realize that everything I had ever seen, heard or felt had already happened, that what I perceived as the "now" was actually already the past!

My experience was always a mental event that followed the actual one. Was there any way to know anything directly... and to know with what? What was this brain or self or consciousness on which these events were being registered? Was the "me" that was asking, the same or different from the "me" I was looking for? Was trying to find my true self a futile exercise like trying to see my retina or trying to taste my tongue?

Quantum physicists claimed that there was no objective reality, that the outcomes of the most rigorous investigations or experiments were influenced by the observer. Einstein said essentially that all phenomena were at the same time matter and energy depending on perspective, that even time was relative. If you traveled a straight line far enough at the right speed you would end up back where you began. That was a terrifying thought to me. My life had not gone all too well to that point. The last thing I wanted to believe was that I could end up back at the beginning living the same frightening moment or even my entire life over again.

No sooner had that thought entered my mind than my fear immediately attached to it and energized it and I soon began to imagine the hundred ways in which I could end up repeating the same frightening and painful life again and again for all eternity.

Maybe it was all just a mental event. The world was just a vision that existed only in my mind, or maybe a much greater mind of which I was only a small part. If everything existed only in consciousness then the phenomenological world, the world of form and matter had to be an illusion of sorts. If matter was solid and real, if that was the basic material of the universe, then where did consciousness come in? When and how did it appear?

If I bought Sister Assumpta's idea that God made me, I still had to ask the question, who made God? Why, oh why was there something, anything… matter, consciousness, even if it was all a dream, why was there anything rather than nothing? I could never quite imagine a beginning to all things, so I set my mind the task of trying to imagine an end. In some ways my fear had become so strong that what I really wanted was to assure myself that somehow, some way there could be an end, a final end to my suffering. I tried to envision matter, decaying down to the atomic level then subatomic then into pure energy but there was always something left from which something new could be reborn. I then tried to follow consciousness down to non existence. I imagined my own small conscious mind losing its sensory faculties, losing any sense of the outside world, closing down its dream factory, its imagination, becoming silent, becoming nothing… and that's when I saw it.

It wasn't the end but the beginning. It was the big bang exactly as Englishman Stephen Hawking would describe it a few years later. It was how nothing became something. In an instant nothing or empty Consciousness or "God" had exploded into Being.

The entire universe had come into existence in a single flash. Potential energy had become energy. Energy implied space and time to realize itself and realize itself it did, literally exploding into what the ancient Chinese referred to as the Ten Thousand Things, or as the term implied, the countless and varied forms in the universe.

Would the stars and all the planets continue traveling on forever into endless space or would the staggering but finite force of the initial explosion only drive the galaxies so far? Would the Ten Thousand Things then follow the laws of physics as explained by Mr. Cost, my ninth grade General Science teacher? Would attraction between masses eventually cause the elements to gravitate back toward each other and ultimately fall back toward the point of origin creating again the gigantic critical mass that had exploded in the first place?

If the universe exploded again, would it take the same shape or a different one? Even if it took form randomly, in eternity wouldn't the same exact world that I was living in be repeated again at some time and wouldn't that mean it would be eventually repeated an infinite number of times. I would have to experience the same unbearable terror and despair without end.

Maybe the reason my terror, my hopelessness was so complete was because my deeper consciousness already knew. It knew this was not the first time I had been here. This was one of many visits to the same hell I had already lived. The indescribable intensity of my fear made me suspect that the unknown part of me that was driving the car when I had drifted off, already knew that this was one of maybe billions of times I had been here before and that there would be trillions and trillions more to come. It was far too horrible to accept. There had to be a way to annihilate myself finally and with certainty. I had to find it.

CHAPTER 3
Bacon

My first word, I am told, was "bacon." As I was raised Catholic and not in a home that kept kosher, it is fairly safe to assume that the word bacon can be taken at face value as a reference to the food item. A person who is in an active phase of Schizophrenia however, might find deeper meanings in even a simple word like bacon. He could wonder for example whether I was announcing that I was a reincarnation of the 13th century friar and scholar, Roger Bacon. The schizophrenic might then begin to look for and find evidence to support that notion in things that would go unnoticed by someone else.

The schizophrenic would find that the machinations I revealed in previous chapters about the nature of reality would be a perfect fit with the Friar Bacon who was known for advocating empirical methods to study both nature and God. It might explain why I questioned my teachers so deeply. The psychotic might even find that Bacon's writings about gunpowder would be a reasonable explanation as to why as a very young boy I found a way to make gunpowder from elements I scavenged and bought from the local pharmacy. The schizophrenic might find more parallels in my life and Roger Bacon's including the fact that we both wrote and so on. The acceptance of these loose associations could reinforce the psychotic's attachment to his theory and driven by his high anxiety to make the world fit his view he would unconsciously look harder and harder to find things to prove his point to himself.

I raise this to make two important points. The first is to show how a schizophrenic can digress into thought patterns which may appear entirely unreasonable or out of touch to others especially if they aren't privileged to the basic hypothesis that the schizophrenic is trying desperately to prove to himself in a futile attempt to resolve his anxiety. From the outside the schizophrenic may seem dull minded and completely unresponsive to his surroundings. Internally his mind could be razor edged and pursuing some elusive theory or thought through a mental labyrinth which the average person could not begin to trace even if they could understand the hypothesis or suspicion the schizophrenic was following.

The second reason is to point out that the tendency to find evidence to support ideas we have already accepted, or want to accept, can be found to a large degree in all of us. Many have preconceptions about how childhood experiences can lead to schizophrenia. A psychiatrist who has many years of training invested in a particular theory of the psychogenesis of schizophrenia could read the following account, finding details that confirm his expectations and fail to see a broader or even a simpler view that might be at odds with his own beliefs. His own beliefs give him comfort and avoid the anxiety of suspecting, say in an extreme case, that his entire approach is misguided, and that he is entirely out of his depth dealing with schizophrenics. How truly different is this mechanism from the one that causes the schizophrenic to desperately seek out evidence to support beliefs that explain or justify feelings that are too strong or frightening to live with? In the cases of both the normal person, as illustrated by the psychiatrist, and the schizophrenic, the extent to which either may miss what is obvious to others may be a function of the degree to which their attention is held inward on what they already believe or want to believe.

As you learn about my childhood and other experiences I had along the way it will be impossible to avoid coming to some conclusions about the nature of my illness and how it came about. Try not to hold too tightly to those notions. My own beliefs about what was happening to me changed many times. In the end, many of my own beliefs about schizophrenia were shattered, finally allowing me to clearly see a way up and out.

I was nine pounds at birth on October 14, 1947 and by all indications healthy. My mother did not breast feed, ever. Her pediatrician, Dr. Aronis, I was told later, did not share Doctor Spock's views on feeding baby on demand. It was up to the parents to set a schedule that fit their own needs and make the little monster adhere to it. That goal was never quite achieved. I had an aversion to milk and my parents complained well into my adult life about how many hours it would take to force me to drink a bottle of it as a baby.

Possibly my earliest conscious memory is of being on the front porch at our home on Riverside Drive in Trenton, New Jersey. It was a narrow semi-detached brick front house on the Delaware River. My travel was limited by the low wooden railing around the fair sized front porch and by the accordion toddler gate that closed off the opening to the four or so brick stairs that led to the sidewalk, the roadway and the river itself. I can clearly remember hearing the clinking of milk bottles just before the white clad milkman appeared from around the corner of the next house. He was carrying a wire frame basket full of fresh milk in one quart bottles along with a couple of empties from next door. I took one of the empty glass milk bottles that were near our front screen door and launched it over the railing at the milkman. He stopped short and looked up startled in my direction as the glass shattered at his feet against the concrete walkway.

As I grinned down with satisfaction at the white suited nuisance I heard the screen door swing wide behind me and slam against the bricks. I felt myself being yanked off my feet by my arm and then swatted on the backside. I don't remember crying. In fact, I can't to this day remember ever crying as a child! My sister, likewise, has no memories of me crying even when I was being hit with a doubled over belt.

I can remember being washed in the kitchen sink and other than that most of the memories I still have from my first three years are about getting away from the home. The neighbor that shared the other half of our little semi-detached home was Pete Pittman. He was a well known sportsman who carved duck decoys and tied fishing flies. I remember going out on the river with Pete in his canoe. The feelings I have that are associated with being on the river are of ease and openness. They are in stark contrast to the feelings I have when I try to remember being in the house. Almost no pictures or images come to me but I can feel a pressure, an internal scrunching, a feeling of holding my insides still.

I have two specific memories of being with my father, briefly. Once he took me into the cellar and I watched as he stoked the furnace with coal from the bin. Another time he took me around the block in a car. My parents didn't own a car for the first couple years of my life. My father came home excited one night, having purchased a beat up blue Dodge that was several years old. He had purchased it from a neighbor, Mike Malcomb, who sold used cars. My father was very proud to have a car and took me for a ride. I remember feeling a closeness or connection to my father that night, much as I did on the occasion when he showed me how he banked the coal stove in the cellar for the night. Those two events may be the closest I ever felt to my father in my entire life. He was usually completely absorbed in work or in following baseball or other sports and left the children to my mother and to her mother who lived with us until she died when I was nine.

As rare as the memories of closeness to my father are I have absolutely none with my mother and I have only the memory of being bathed in the sink and a vague memory of being in the parlor with my grandma to show for the three years in the house on Riverside Drive. I do remember being outside on the porch, on the sidewalk or on the riverbank and some of those memories include other kids in the neighborhood. I know from pictures and from one vague memory of a struggle with my mother that I was walked on a leash that was connected to a white leather harness I was strapped into.

I do have a vivid and complete memory of leaving the little home on Riverside Drive for the last time at age three. It was dark out and it was raining quite hard. My mother ran to a car parked at the curb carrying my one-year-old baby sister, Barbara. My Grandma was with us. The rain was pounding so heavily on the metal roof of the car that my mother and the driver had to shout to be heard. The wipers were slapping frantically side to side but hardly making any difference. A lightning flash briefly lit the determined face of our young driver as he leaned forward and strained to see through the rivulets running down the windscreen. I clearly remember thinking how brave this man must be.

We were moving to the San Francisco area where my father had been transferred and promoted from sales trainee to field sales agent for the De Laval Steam Turbine Co. My father had driven out six weeks ahead of us in the oil burning blue rattletrap. We were being driven from Trenton to LaGuardia Airport in New York by a couple that was friends with my mother, Bob and Ann Chamberlain.

The storm was extreme. The thunder and the lightning boomed and flashed continuously. The lower limbs of the big oak in the middle of the intersection at Lee Avenue waved wildly at us as the car crept along through the deep puddles and the streams of runoff racing toward the swollen river. As we rounded the corner I turned and strained to look out the window, not back at the little half a house which we were leaving forever but out across the open expanse of my beautiful river which flickered with a deep darkness and then a shimmering silver under the lightning filled sky. The river sped along with a purpose, ignoring the wind and pounding rain, as if it knew where it was headed.

That night, possibly at the very moment I had seen the lightning illuminate the determined face of our brave driver, something in me changed or matured. My awareness or consciousness became more complete. From that night forward my memories of childhood become lucid and complete and the gaps are few.

I remember much of the airplane flight to the west coast. We were on United Airlines and the stewardess gave me wings to wear. As that memory surfaces I am struck by the completeness of it. As I write this I can feel the sharpness of the pin which fastened the wings and even the coolness of the wings which reminds me that they would have been made from metal in those days, rather than plastic in more recent years.

We were on a two engine propeller plane, possibly a DC-3 or 4. My grandmother was nervous. I am sure it was the first time any of us had flown. She stopped a stewardess and anxiously told her there was fire coming out of the engine on the wing outside her window. The stewardess told Grandma to be sure to let her know if the fire stopped. "Then we might have a problem," she told grandma reassuringly. We made three stops along the way to take on fuel and food. The stewardess gave me the captain's hat and filled it with chewing gum to pass out to the other passengers as we descended. This was to help with the ear popping from altitude change although I had no idea of that at the time. I was merely delighted with the attention and affection I received from the stewardesses, which on reflection makes me sad to think what a rare and unknown feeling that was to me at that time.

CHAPTER 4
California

When I think back to my time on Riverside Drive in Trenton, my insides scrunch up as if I were holding myself inward. I feel tension in my ribcage and my stomach contracts. I seem to be holding my head still and looking warily around with only my eyes. Someone once told me that you can spot people who have been in prison by the way they scan a place with their eyes, moving their heads only slightly and their bodies not at all.

When I think back on the first year or so in Castro Valley, California, the feelings are quite different. I feel open and at ease. Again I have many more memories of being outside than memories from inside the tiny house. I played in the backyard with Eddie Kramer who lived next door. My family did not own a television nor did Eddie's family and I don't think either of us had a lot of toys so we just dug around in the dirt and built things with sticks and stones we had found.

Our street ended at the foot of a good sized hill that was covered in the golden dried out grass that is prevalent in that part of California in the dry season. Some afternoons there were a dozen kids gleefully sliding down the hill on flattened out pieces of cardboard cartons. I was told it was too dangerous. Staying in the house was presumably safer although one day crawling around the floor I found a good sized rusty nail and exploring further found some holes in the wall to poke it into, just to see what was in them. What was in them was 120 volts of electricity which shot up my arm and into my head with a buzzing rage that knocked me flat. My mother took the nail away from me less ceremoniously and with less interest than when she had told me I couldn't go hill sliding at the end of the street. "That's not yours."

At the other end of the street across the boulevard was a subdivision of little wooden and stucco homes much like the tiny postwar bungalow that we lived in on Tyee Street. The homes in that neighborhood were all abandoned in spite of the fact they had only been built between the end of the war around 1945 and 1950 when we moved there. The postwar boom in the suburban areas caused congestion on the roads leading to and from the cities. Now the government was building wide new "freeways" to connect these areas and was tearing down the recently built communities to make room for the roads that were being built to serve them.

Many of the dozens of vacant little family homes had been broken free from their foundations and were resting on three or four-foot-high wooden jacks or pilings. Occasionally a crew would load one of the homes onto a big wide flatbed truck and haul it off, creaking and groaning, down the boulevard to be affixed to a new foundation in another quiet little area so the roads there could become congested.

One day a couple of the older kids came and sat out on the front stoop and showed me a cigar box they had filled with books and boxes of paper and wooden stick matches. When my mother came out they told her they had been collecting them and asked if she had any old ones she could give them. She told them no and went back inside taking me with her, while muttering something about what eight year olds were doing with matches.

Apparently she didn't think to ask the neighbor kids' parents the same question which might have avoided what happened two weeks later. Once the kids decided they had enough matches they made their way across the boulevard, packed the wooden pilings under the abandoned houses with multiple match packs and ignited them. Several homes were destroyed by fire. When questioned about why they had done such a thing, the kids explained simply that they had been learning all about the fire department at school but had never actually had a chance to see the fire trucks come to a fire. Lighting a number of houses on fire they had decided unanimously would be a great way to see all of the different types of the trucks in action, not just one or two.

The other big event that year was at the Noloroths' house. They lived across the street and a couple of doors down at the corner. My family and a couple of others including the Kramers were invited over after dinner one evening to witness a blessed event. The Noloroths had purchased a television! The television screen itself was almost round. As small as it was, it looked even smaller compared to the big wooden cabinet it peered out from. The cabinet looked much like the grand old mahogany radio console it was replacing. The TV was placed in the corner of the room where the radio had been and was nearly obscured by the overstuffed arms of the chair and sofa that it was nestled between. The television was thought of more of less as an upscale radio for the richer folks and no one had yet thought to orient furniture toward it, even less to make it the center of attention in a room. To see it, most of us had to sit on the floor looking back toward it and the main furniture pieces in the room. The picture was fuzzy with "snow" and Mrs. Noloroth had to keep instructing Mr. Noloroth to jump up and delicately turn little knobs to get rid of the diagonal squiggly lines that kept coming back. Between adjustments we got a look at black and white images of a cowboy in a great big white hat romping around the countryside on a horse. We didn't stay very long. Although no one said so, most everyone was pretty bored except for Mr. and Mrs. Noloroth and their son Gary who lay on the rug on his belly, elbows bent, chin in his hands, apparently frozen in place by the flickering image.

Sometime after I turned four, three things happened that undercut the sense of ease and safety I initially had in California. I don't remember which order the three events happened in so I don't know whether one event precipitated or contributed to another one but I am certain that the three events did compound one upon the other to kindle in me, quite a different sense of my world.

I went off to Baywood Elementary School happily and eagerly. I was still four and one of the youngest kindergarteners because of my October birthday. My next door neighbor Eddie Kramer was my best and only friend and we were dropped off together the first day, excited about sharing our new experience. Within the first few minutes while we were still standing outside the classroom door, one of the school bureaucrats noticed that Eddie and I knew each other and instantly and without discussion or explanation separated us sending Eddie to the classroom next door. The purpose behind this I can only guess at. Maybe they were afraid we'd band together and recruit the other students into one of the early Oakland area street gangs although it is likely they believed we'd be more open to socialize and make new friends if we were on our own.

How'd that work for us? To this day I can't remember a single name or face of any other kid in the class. I did not have much of a clue about most of the things we were being taught or exposed to but interestingly the memories I do have are of working with physical things like clay or paint rather than with symbols, like letters or numbers. We did begin to read about Dick and Jane and Spot and that all went smoothly. I think I could read as much by associating the pictures in the book with the spoken words as by remembering the written symbols or letter groupings.

There was a jungle gym in the classroom's own separate play yard. Eddie's class had its own play yard next to ours but separated by a six-foot sapling fence. Eddie's yard also had a jungle gym and at recess one day we found that we had each climbed to the very top of our respective jungle gyms to escape the mayhem below and that we could talk to each other across the fence. From that day forward we spent most of our time at recess talking to each other across the fence from our perches above the crowd. Eddie and I walked home from school together on occasion but in the mornings I usually walked the several blocks by myself.

Our family had one car which my father always took to work across the bay in San Francisco. The school was a few blocks away and the route had only three or four turns. After the first day of school I was left to find my way back and forth alone. There was a specific day when this seemed strange to me and when I remember feeling vulnerable and anxious about it. It was drizzling rain and was grayer than usual. I walked by a small general store on the way up the boulevard which was no more than a two lane street with houses and a small shop or two every couple blocks. There was a dark grayish green sedan outside the little local grocery store and there was a man hunched down in the driver's seat. I had the feeling he was watching me without turning his head.

A feeling came over me that I could not have described at the time and which I have a great deal of difficulty describing or understanding even now. I remember thinking to myself, "How could my parents let a little kid like me walk alone here where somebody like this man might do something to me?" The feeling that came with that thought is not easily explained but it was a dirty, violated feeling. I have no memory of having been physically molested at or before that age yet even now sixty years later I can't help but believe that the man in that car wanted to do something to me. It was almost as if when I walked by the store that day I passed through a section of polluted mental space for want of a better description. The dirty feelings, the feelings of being violated somehow stayed with me in my groin area and also around my face. That feeling that I was somehow vulnerable to that man in the car, the dirty feelings, and the feelings that I wasn't being protected came back often. They were vague and kids are not taught to pay attention to their feelings so our attention is not on them. I found an alternate and longer way home from school that afternoon and often took it to avoid walking by the little store where I first had the feelings.

The second event that was a factor in my increasing discomfort happened as I lay in bed at night. My legs would start to feel somewhat numb and very heavy. They felt like they were very long and my feet felt very far away. I could wiggle my feet and see that they were right there under the blanket in the tiny bed but they seemed much farther away. The heaviness was extreme and there was a feeling of darkness in the lower part of me.

When I would shut my eyes to try and sleep the heaviness would increase and I would sometimes feel as if my legs were being pulled or distorted from all sides. I had a vague sense of some commotion going on out beyond my feet, between where my body was and some dark hills close by. The area around the hills and the "other" or distorted body seemed a part of some parallel dimension or netherworld to which I was partially connected. I would lie awake for what must have been hours. When I did sleep, I would dream the same feelings of my legs being long and heavy and distorted. In the dreams I could hear a sound like a vacuum cleaner and my legs would feel as if they were being distorted out of shape, pulled by the vacuum cleaner although it was invisible, somewhere off to my left. Even at the age of four I knew that this was not usual but I did not speak to anyone about it any more than I spoke of the feeling of violation that I had walking to school.

One night I became aware of my parents and a doctor standing over me and they were all quite upset. This was the third event. I had contracted a severe case of the mumps, what my mother and the doctors later claimed was the worst they had ever seen. I had convulsions and "near brain meningitis." The doctor had given a sulpha drug to fight the fever and infection. I had a reaction and my heart stopped; more injections, probably adrenaline to keep me going. I was soon placed in an old fashioned ambulance on its way to Children's Hospital in Oakland. The doctor rode in the ambulance, massaging my legs to keep the blood circulation going. I could hear the wailing siren and when I bent my head back to look upside down through the windshield I could see the ambulance was running red lights. I knew this was a big deal.

I was put into a quarantine room… alone. I was in a high sided metal crib. The room had peeling greenish paint and a heavy yellowed door. There was a small square window in the door that had wire mesh imbedded in it. I was angry about being left alone. The room felt dirty. The peeling paint was bad enough but this was dirty space. Again for lack of a better description, the space itself felt dirty. This wasn't dirty in exactly the same sense as the space I had passed though near the man in the car. This was dirty with sickness. Occasionally the head of a nurse or someone else would pass by the little wired window and only once did one of them acknowledge me when she put her head in and asked if I wanted a ginger ale. "No!" I shouted back at her, completely unable to communicate the abandonment, frustration, discomfort and psychic contamination I was feeling.

CHAPTER 5
'Lectricity!

When I was still five we moved to Seattle. My father had been promoted to sales manager for that area. We went by car and I saw my first waterfall along the way. It was just a glimpse as we passed by. It was off to the right and fell from quite a height into a stream that ran along next to and then under the roadway. Just the brief sight of it instantly brought back a sense of ease and lightness in me.

We stopped overnight somewhere in Oregon at a small roadside motel. My two years younger sister Barbara accidentally locked herself in the bathroom. I had several anxious moments listening to my sister's frightened pleas and cries while my father wisecracked that we might have to leave her behind to stay on schedule. Eventually an employee showed up with some kind of tool that went into a hole in the handle and freed my "sisser baby." I think I was as relieved as she was.

We moved into a two story home near 4th St. NW and McGraw Avenue on Queen Anne Hill. I was still in kindergarten and picked up mid year at Coe School a few blocks away. The kindergarten classroom was a story above ground level and looked out over the playground much the same as in the movie Christmas Story. I spent a fair amount of time stretching up on my toes to look outside where I often wished I was. There were quite a number of toys and other materials in the classroom including a set of giant wooden building blocks. Many of the blocks were a foot or so, on a side. When we were allowed, I loved to help build ships or firehouses or anything else we could imagine. I don't remember doing much of anything else except lying on the floor and pretending to nap when we were told to.

Outside at recess one day a mean little kid came around with a tube of some kind that was oozing metallic goo, which I much later came to believe was a lead containing product known as "Liquid Solder." He was cornering individual girls and boys against the wall of the school or in one of the corners of the playground fence, squirting the metallic goo at them and screaming "LECTRICITY" at the frightened children. There was a shocking viciousness and intensity in him that I had never seen before. He had accumulated a hoard of followers who howled along with laughter as each new victim was cornered and confronted, then ran broken and screaming in fear. I looked on frozen in horror as the mean kid backed a little girl in a plaid jumper against the metal grate over a window in the wall of the school. He thrust the oozing tube into her face and screamed "LECTRICITY" at her as she collapsed onto the ground crying and covering her head with her arms.

Just a moment later the bell rang and we began to gather inside the school entrance on the stairs that led to the upper floors. Over my shoulder I caught a glimpse of the mean kid making his way through the crowd behind me and before I could flinch out of the way he had thrust the ugly ooze into my face and with his beady little rat eyes focused in on my face to watch my fear he breathed his "LECTRICITY" on me with all the evil venom he had in him. I raised my arms, clenched my fists and began to pound down on his head and shoulders like an out of balance windmill running at triple speed. The most surprising thing about the whole affair was the surprised look on the imp's face as he staggered backwards falling against the students below him on the stairs.

Where the teachers had been during the entire recess when little Attila and his Huns had been terrorizing their dozens of victims, I will never know but I do know that each and every member of the entire faculty at Coe school was standing within a half step of me at the very instant I began meeting out corporal punishment to the little demon. "HEY, HEY, HEY!" they yelled out in unison as several of them caught hold of me and dragged me up the stairs, while the rest of them bent down to hold and comfort the poor little victim.

We still did not own a television and I don't remember having seen a movie at that point in my life. I don't believe I had ever seen a fight and I have no idea where I got the idea to make fists or to hit someone with them. That was the only time I got into a fight that year or even in my whole life until I entered Catholic school. The incident was never discussed with me by the teacher or anyone else yet the only comment on my report card that year was

"Tom solves his problems with hard fist blows."

This kind of incident again increased the distance in me between what I felt to be right or true and what I came to believe could be spoken about or acted on in the world outside. The next door neighbor Patrick Harris was about four years older than me. He sometimes played with me in the rock garden in front of our house. He had a number of small toy cars and we would make little roads for them among the rocks and plants. One day he lent me two or three of the older beat up ones. There was a little black rubber delivery truck and another small car or two. I was delighted and thanked him. I took them inside and played with them in my room and even drove them around the rim of the tub when I was in the bath.

Three or four days later, I was in the front yard with Patrick when he complained quite angrily that someone had stolen some of his toy cars and when he found out who, he was going to beat them up like they had never been beaten up before. I started to open my mouth to remind him of the obvious, but I stopped myself. What stopped me was the same vicious intensity in Patrick that I had seen in the little mean kid with the "lectricity" goo. I couldn't believe he had already forgotten he lent me the cars but I froze. I was afraid if he was stupid enough to have really forgotten he lent me his toys, that he might not believe it when I told him the truth and I would catch a beating anyway. I hadn't had a real beating at that point; even most of what I got at home was mild until we moved back to New Jersey. I had no real memories of being hit like he was threatening but I was truly afraid of the meanness and intensity of his threats. Once again, I learned to ignore what I knew was right and true and keep it inside. I hung onto the little cars for some time thinking Patrick would soon remember he had lent them to me and I could give them back to him. When that never happened I simply snuck them into the trash, thinking that would put an end to the problem.

Years after Patrick had likely forgotten he ever had the cars; I still obsessed over the incident. It bothered me that I had never found a way to return the cars. It bothered me that I had essentially been untruthful, and it bothered me that it bothered me, which is to say that it upset me to think that Patrick's bad memory and meanness had caused me to be afraid. It wasn't right but I didn't know what to do about it and I apparently didn't feel there was anybody I could talk to that would understand.

CHAPTER 6
Dr. Silverston

In a more extreme form, the problem of believing there is no one you can talk to that could understand… is paranoia. During my breakdown in the early 1970's, on my worst days I believed that no one could understand what went on inside me. Beyond that, I also believed that there was literally no one out there to talk to at all. Some days it would seem to me that the people I encountered weren't really people but just a veneer of sorts, a kind of ultra thin plastic like surface with nothing behind it. There were no organs inside; no flesh or bones behind the facade. If I could somehow sneak a look behind I would see the people didn't even have backs. They were like the movie sets in westerns where the little frontier town is made of buildings that have fronts with nothing behind them. There wasn't anything concrete in the world. It was entirely a kind of illusory projection not unlike movies or dreams in which we accept what's going on in front of us as real until the movie ends or until we awake.

The big question for me was whether the projections were coming from some unknown part of my own mind or from some inconceivably evil spirit entity. To speak my inner thoughts, my suspicions, aloud to one of the fake people could be to speak them directly to the evil thing, the devil that had created the entire illusory world to trap me into the eternal and empty void. On the other hand, to fully accept that the illusions were coming from me, that the world was actually my own projection, was to accept the void as the ultimate reality. If there was nothing in the world except what was in my mind, only me and no true "other," then ultimately I would have to accept my aloneness for all eternity.

The terror that came with these thoughts is probably unimaginable to most people who haven't experienced it. I had suffered through many, many severe beatings, near fatal illnesses, a prison riot and a war. During the war there were many horror filled days and nights including a time when I and only one other Military Policeman came up against a full platoon of thirty hardened Viet Cong soldiers in a firefight. The fear I had experienced before the schizophrenia was virtually insignificant compared to the searing, blinding, constant terror that haunted me day and night during my breakdown. Walking along a street I would suddenly be overtaken by the feeling that the ground beneath me was about to dissolve and I would fall into an eternal pit and burn forever in the black flames of hellfire.

I was caught in a trap. The trap was the phenomenological world which was just an illusion whose sole purpose was to torment me and trick me into worse and worse hells. If I shared any of this with anyone, I could merely be giving information about my limited defenses right to the enemy, to the devil that was behind it all.

Much of the time I fought fiercely to make the world real, but the nagging fears and suspicions were always there to undercut my progress. I turned the radio on in the car. The disc jockey, Edward Simpson from the hip FM rock station in Philadelphia, was taunting me. His spooky low tones and esoteric jibes were so specific and personal that only I could understand. If I had tried to point it out to someone else, I knew they would think I was imagining things.

There were moments when I had hope, however faint. Something would seem too real or complex for me to have dreamed up. Sometimes I would interact with someone, like my old high school friend David, someone who had a soul and a mind of his own, unlike the great majority who were mere automatons controlled by the big evil mind in the sky.

In one of those fleeting moments of hope in July of 1970, I decided to see a psychiatrist, Dr. Melvin Silverston. His name had come from my Aunt Eileen, who like all of my several aunts wasn't an aunt at all but a friend of my mother's. Mom said a friend of Aunt Eileen had seen the doctor briefly when she was having a problem. Aunt Eileen was referred to by some of the other "aunts" as Crazy Eileen. She, like my mother, had been born into a well off family and treated like a princess until both families lost every nickel in the depression. Like my mother, Eileen still thought she was royalty but unlike my mother, Eileen still found a way to show some of the trappings of wealth even though she could obviously ill afford them. She drove a long black Chrysler Imperial and smoked slim little custom cigarettes through a fancy jeweled cigarette holder. She wore rhinestone studded cat eye glasses, bracelets and necklaces that all accentuated her Auntie Mame affectations. She dressed her black miniature poodle, Pepe, in yellow rainwear and bright red rubber booties.

I assumed that Crazy Aunt Eileen's "friend" was an imaginary one made up to conceal either her own trips to the headshrinker or those of her son, who might have been in need of help simply from living with his mother and Pepe the poodle. Either way I was certain Eileen's recommendation was a more direct one than she let on. A referral from Aunt Eileen for a head doctor wasn't exactly a ringing endorsement, but I was truly and finally desperate.

There was no receptionist at Dr. Silverston's office. I sat alone in the waiting room. It was quite dark. The walls were paneled with dark wood. The carpeting and the furniture were dark and the lighting was subdued. The only bright spot in the room was the top lit aquarium across from me.

I assumed the dark surfaces and the low lighting were to provide a low stimulus environment to help the patients relax. That might have worked for me except for the big blue fish with the yellow mouth that kept staring at me from the corner of the fish tank. It occurred to me that the fish could be being used as a spy; its eyes sending information on me back to what ever devil's disciple had sway over this portion of creation. At that point in my sickness I would not have been completely surprised if Satan himself had opened the inner door and invited me in to his office.

Pushing those thoughts aside, I gutted out the next few minutes until the door finally did open. As long as I had waited, I expected to see another patient emerge and I had hoped to gauge something about the coming session from his or her demeanor. No patient came out and instead a man in a dark suit and tie greeted me. I was immediately relieved to see he didn't sport a Luciferian goatee or even the more expected beard of the Neo-Freudians.

Seeing the doctor, for the moment began to ease some of my fear that I had been dreaming up reality on a moment to moment basis. I could not have dreamed up Dr. Silverston in a month of Sundays. The blank stare he directed at me was similar to the one on the face of the big blue and yellow tropical fish treading water in the outer office. The inner office was as dark as the waiting room. There was a big bookcase filled with leather bound books, a wall filled with diplomas and a tape recorder on the desk. I thought a neatly trimmed psychoanalyst's beard might have made a good addition to the other accoutrements of the psychiatrist's trade because in his case it might have hidden the fact that he looked pretty much as anxious as I felt.

The doctor did not take a formal history of any kind. I told him how I had begun to doubt my perceptions and how even at that moment I could not be sure whether I was awake or dreaming or trapped in some parallel reality. I told him how I had become overwhelmed by my feelings of despair and especially by the attacks of terror.

He stopped me in far less than the half an hour allotted for my appointment and quite cavalierly diagnosed my problem as alcoholism. I told him I had drunk a good bit since the war and in the last few months had even augmented that with some of the marijuana and related substances that my old high school friends always had with them, but in recent weeks I had not had much of anything at all. He reassuringly handed me a prescription for a tranquillizer, "Trust me," he said. "This will help you to feel better."

I did feel a little better as I left the office. Just confessing my fears had probably taken some of the power from them. I was confused about how alcohol use could have anything to do with what I was experiencing but I was desperately in need of something to hope for. If taking a tranquilizer would help, I would certainly try it. I drove the few blocks to Adams and Sickles Pharmacy. It was an old neighborhood drug store that was owned and run by my friend David's dad. What Dave's dad, pharmacist Jack Harvey, knew instantly but did not share with me, was that Dr. Silverston had lied to me. If he believed I was alcoholic, that wasn't all he believed. The drug he gave was a far cry from the minor tranquilizer he said he had prescribed. It was a powerful antipsychotic drug named Thorazine. I believe the dose was 50 mg. two or three times daily. Typical recommended starting doses are as low as 25mg once daily. The highest dose recommended is 100mg. three times a day except in "rare cases in a controlled setting when administered by highly specialized doctors."

Thorazine, or chlorpromazine as is it is known generically, reduces neural activity by blocking certain dopamine receptors and post synaptic sites. It is a part of a class of drugs that are phenothiazine derivatives. The more basic compound is also used in insecticides, killing bugs by the same basic function of blocking nerve impulses.

Soon after taking the first pill I developed a very pronounced sense of butterflies in my stomach. The feeling was fairly constant and quite uncomfortable. It did counteract or at least distract me from the sense that my body was about to dissolve into space, which often triggered my sudden feelings of terror and the obsessive thoughts which followed. I also became quite tired over the next couple of days.

My sense of dread was still at the edge of my thoughts but the tiredness and the stomach discomfort kept me from following my mind as far down the panic path. I tried to convince myself I was feeling better. A couple of days later I decided to try to resume my normal activities. I had been heavily involved in car building and drag racing since my mid-teens. I decided to travel to York, Pennsylvania with my race car partner, Ed Franks and the rest of our volunteer crew, to race our national record holding 1957 Chevrolet station wagon, known as the Cherokee.

In high school I was a "wiz kid" in Power Mechanics class and my instructor Mr. Hornchek got me a job installing aftermarket automotive air conditioners immediately after graduation. When my boss, Ed Krelman, was decapitated in a car accident, his father in law, Henry Schwartzburg, got me an interview with the service manger at the local Pontiac agency.

I started at Cathcart Pontiac as an apprentice mechanic and lot boy and was again soon recognized as highly motivated and unusually talented with cars. I was invited to attend General Motors own trade school for mechanics. I went at no cost to me and in fact I was paid while I attended. Almost immediately after I graduated I was promoted to New Car Service Manager and at only age eighteen I proudly handled new car prep and deliveries as well as all regular service and repairs on cars up to their fourth service visit. I supervised mechanics more than twice my age, including dealing with one with an alcohol problem and two others who I caught defrauding the system by charging for warranty repairs that they never made. I interacted directly with customers and salesmen and was trusted to resolve problems without supervision. I was especially good at diagnostic work and had a special genius for tracking down electrical problems. Although there were mechanics in the shop that in some cases had thirty years on the line, I was especially proud of the fact that I was the only employee who could diagnose and repair both automotive air conditioners and automatic transmissions.

Near the end of the summer of 1966, with the help of "The Good Guys" at the Trenton Speed Shop I built a motor for the Cherokee '57 Chevy station wagon of my new partners, instantly elevating the car to a real contender both locally and regionally. By the time of my visit to Dr. Silverston in 1970 we held six National Hot Rod Association national records in three different racing classes.

Even though I still had the butterflies and periods of tiredness, we towed the Cherokee to York, Pennsylvania to compete in a three-day regional drag racing meet. I continued to take my medicine. The obsessive thoughts and the feelings of unreality came and went but with somewhat less force. Eddie, who was half Cherokee Indian and my partner, drove the race car and I performed my function of helping to tune and retune suspension, drive train and engine parameters to maximize performance under changing conditions of air and track temperature, engine temperature, air density and other factors. We changed the rear end gears, ignition advance curve timing, adjusted carburetor jetting, and manipulated tire pressures. We kept a log book of hundreds of prior runs under varying conditions at different tracks with different concrete and macadam surfaces. A combination of calculations and guesswork based on gut feel told us which setup to use on a particular run… and we were quite good at it, which is why we held national records.

During my first trip to the annual national drag racing event near Indianapolis, Indiana a number of well known racers from the North Eastern Division gathered around to hear my take on a different way to calculate carburetor jetting to achieve optimal fuel air mixture for particular combinations of air density and air and engine temperatures. I was eighteen and proud to be taken seriously by my older peers. Now at twenty-two and a half after being away in the Army for three years, I was a half year back into racing. Although at moments I was struggling to keep from feeling lost forever, my sense of reality that first day of the meet at York was strong enough to keep me engaged most of the time and I was doing my part to keep the car performing well.

That night we camped out in the pits at the track and as the sun was going down my skin began to burn. I felt as if I had severe sunburn but I could not see any redness or feel any blistering. As the evening cooled off the burning became more severe. I was unable to sleep for most of the night. We kept large quantities of ice with us to cool down the intake manifold, fuel line and radiator between runs. I began to take ice from the coolers and I spent much of the night sitting up in my sleeping bag rubbing it on my arms and face.

In the morning the sensation only increased. Two hours after the sun came up my forearms and face felt like someone had taken a propane torch to me but still, I could find no external evidence of sunburn. Icing failed to give me anything but momentary relief and by midday I took to crawling under our tow vehicle to find a shady spot out of the sun.

My mind searched for an answer to how I could seem to be sunburned but show no signs of it. Suddenly and with a pang of terror the answer hit me. I had been right all along. I was in Hell. I had tried to deny it. I had taken the pills the doctor prescribed to hide from it. There was no escape.

As I had suspected if I tried to kill my self, to escape by taking drugs or even to pretend the horror didn't exist… it would only make it worse. The nuns had been right. They had said we would burn for eternity. I didn't see how at the time. It didn't make sense to me then. I couldn't see how someone could burn for any length of time without being consumed and destroyed by the flames… but I could see it now. I was living what they were talking about. My skin was burning without being damaged. The hellfire was inside. I knew it would soon get worse and I would want to kill myself. It wouldn't help. I was already dead. This was just a dream that would get more hellish if I tried to escape. It was cruel and terrible but there was no escape.

I spent the next day and a half hiding under the truck we used to tow the race car, rubbing ice on my skin. Riding along the Pennsylvania Turnpike Sunday evening, I kept hanging my head and arms out the passenger side window trying to get what relief I could from the wind. The burning was severe and the fear overwhelming. I thought constantly for almost two hours about opening the door and throwing myself out into the roadway, but I knew that would not end my suffering. After another couple of tormented days and nights I ended up back in Dr. Silverston's office for my second appointment.

I was more convinced now than ever that my situation was hopeless. If the office had seemed dark before, now it was far darker. My fear and distrust of the doctor was increased. I said little and instead watched him carefully for hints as to his true intent. It occurred to me that he might not himself be aware of his purpose regarding me. His somewhat distracted or disinterested manner suggested that he was only partly present to the situation. The great evil could have invaded his being and been manipulating him to achieve its purpose against me. There was a disturbing darkness in the undercurrent of his being that I had only noticed peripherally during the first visit.

I had been so very desperate, so needful of hope of any kind that I had overlooked the very anxious, quietly desperate man hiding behind his desk and diplomas. It was possible; I thought to myself, that he might be as desperate as I am. He might be trapped as I am by the darkness and afraid not to do its bidding for fear his own situation would worsen.

I winced as I tried to rub away the burning. He questioned me about it. "Oh, that's merely photo sensitivity. It's just a minor side effect from the tranquilizer, just like the dryness in your mouth." His offhand attitude was dumbfounding. I sat speechless as my mind reeled back through the thoughts of suicide and damnation that had run rampant over me for the last several days.

He quickly grabbed for his prescription pad. “Here, take this instead.” I reached out for the slip of paper on which he had written a prescription for Mellaril. I remember it as being a very high dose, perhaps as high as 200mg. to be taken “QID” or four times a day. Recommended starting doses range from 50mg. to 100 mg. three times a day to be gradually increased later, if necessary to a maximum of 200mg. to 800mg. a day, divided into four doses. It was a drug that would not only fail to help me but would seriously worsen my condition and create serious side effects that would quickly land me in the hospital.

CHAPTER 7
Beatings

Just how out of his depth Dr. Melvin Silverston was in trying to deal with someone as deeply disturbed as I was would surface in less than a week when horrendous side effects from the new drug would devastate me. My parents would call him in a panic as they saw my condition implode. The good doctor would say he would not be able to see me anymore as he was about to leave on vacation.

"Oh, he goes on vacation every time he gets a patient he doesn't know what to do with," said the receptionist at the new psychiatrist's office. "Half of our practice comes from referrals of patients that Dr. Silverston can't handle," she told my mother cheerily.

I wasn't sorry that I didn't have to see Silverston again. My friend David's father, the pharmacist, had asked Dave about what was going on with me that I was on an antipsychotic drug. When David shared that information with me it came as a shock. It wasn't so much a shock to realize that the doctor had believed I was psychotic although the word schizophrenic can still send a chill through me. The bigger shock came from the realization that the doctor had been boldly dishonest with me. How out of touch had he been to think that I could take a drug like Thorazine without soon learning its purpose? Why hadn't he prepared me for that? Even worse, he had failed to alert me to the potential side effects that had caused me anxiety and fear that nearly drove me to suicide.

We are conditioned to assume that doctors are all highly educated, up on recent advances in medicine and especially competent to heal in their chosen specialties, like psychiatry. In the late sixties and early seventies there were doctors still practicing that did not have several years of medical education. Some had never even been to college. Before Carnegie grants to the medical schools restructured medical education there were doctors who had only studied six months before beginning to practice. Some medical schools in the thirties had only one cadaver and not even a microscope.

Something had bothered me about Dr. Silverston initially but because of my desperation and because of the deference we are taught to afford doctors, I overlooked my intuition. There was something incongruent about him. I had a sense that he wasn't completely present to me. He wasn't really there as the person he truly was but as someone playing a role. He was playing doctor as it were and if I had been able to attend more fully to my feelings about him, I would have had to conclude it was a role for which he was not entirely fit. On reflection, I could say the same problem existed in me. I wasn't really being me, questioning the doctor directly and more fully about how alcoholism could cause my condition or about any of the other concerns I had. Instead I was playing "patient" deferring to the doctor based on a role I had been taught.

It is that very split between who I was inside and who I came to think I should be that was likely at the heart of my disease. There were many factors that kept me from growing into the fully actualized person that I may have been capable of being. The beatings that I suffered as a child were most certainly a major part of the picture.

There are some things I have experienced that I know I will never be able to completely communicate. Among them are the level of fear and terror that I lived with for years during my schizophrenia and the uncompromising grief and sadness that I carried home with me from the Vietnam War. These things are simply so far beyond most people's experience that whatever words I may find to relate the feelings are only approximations. The viciousness and relentlessness of the beatings I suffered through for years as a child is another of those experiences that are so foreign to most people that I can probably only hope to portray a small glimpse of what it is like to live that reality day after day, yet it is a most crucial part of my story.

I was in the second grade in 1954 when my family moved back to New Jersey from the West Coast. We moved into a rented house on Parkway Avenue in the quiet Trenton suburb of Ewing Township. It was a two story flagstone covered Dutch Colonial owned by our next door neighbor Frank Naples, who was Chief of the County Detectives. I became ill with the third and fourth of my serious bouts with the childhood diseases shortly after moving in. In Castro Valley California I had nearly died from the mumps and in Seattle I had had an extreme case of German measles. On Parkway Avenue I had a different strain of measles followed by a very serious case of chicken pox. According to my mother, I ended up weighing only twenty-seven pounds when even the smaller boys in the same grade weighed double.

A year later, at the beginning of the third grade, we moved again into a nearby neighborhood where I became subject to daily beatings by three bullies from the Catholic school I attended. My parents purchased a home, also a two story colonial, just two blocks away on Latona Avenue. I was given my own room, a large dormer room over the garage that was cold in the winter and way too hot in the summer. Air conditioning was as yet uncommon, and in our world unheard of. I had a great deal of trouble falling asleep during those years on Latona Avenue. I would lie awake for many hours each night thinking about escaping down the river on a raft like Huck or taking out the bad guys in a gunfight like Wyatt and Doc alongside Virgil and Morgan. Those many hours awake and the wildness of my imagination were fueled by the adrenalin that flooded my system due to the physical beatings I was suffering at the hands of the three bullies during the daytime. Those nightly flights far off into my mind were among the very few moments of peace I knew during those truly terrible years.

The pressure on any kid attending Blessed Sacrament Catholic School was extreme. The days were highly structured and the workload was enormous. The nuns ruled by wielding threats to both our physical bodies and to our eternal souls. Sister Frances Theresa, a very short very round woman who looked not unlike Friar Tuck from the Robin Hood movies, carried a leather wrapped blackjack concealed somewhere within the folds or under the front piece of her habit. Sister sold penny candy at recess from a cigarette girl like tray or shallow metal box that was cantilevered against her substantial midsection by a strap around her neck. Apparently Sister lived in fear of being rolled for the proceeds from her penny candy sales and so carried the sap which she would sometimes heft to back off any fourth graders who got so close to the tray that she had to worry about them pinching a licorice stick. Sister Assumpta used a yardstick or a chalkboard pointer as a swagger stick inside the classroom and on occasion in the play yard. Miss Heitzman, the only lay teacher at Blessed Sacrament, was armed with an icy and piercing stare that could freeze an offender in his tracks quicker than Sister Frances Theresa's blackjack.

The nuns also had few reservations and apparently no conscience at all about using humiliation as a weapon. On more than one occasion as a third grader I was made to stand in line with the first grade girls as punishment for some minor and unintended infraction of the all encompassing and ever more detailed rules. One day in fifth grade David Butler raised his hand along with the rest of the class indicating he had seen the television program we had been assigned to watch the night before. The nun called him out and made him stand while she harangued him. Butler came from a very poor broken home. He wore the same clothes every day and went for very long periods of time without a haircut. "What? Whaaaaat? You have a television, Mr. Butler? You can't even pay for your books and you have a television? You go right home tonight and tell your mother I want to see her. Sister continued to mumble to her self as she stormed over to her desk and made a note to herself while David Butler slinked back down into his seat and bowed his head in shame. It was never certain whether the Butlers or several of the others really did have televisions or whether like me they had raised their hands because they were afraid not to.

Even the kids who were not being beaten up were often afraid to go to school and the tension was palpable at the bus stop, boys checking to make sure their neckties were snug, girls fearfully comparing answers on their homework assignments hoping they wouldn't be singled out as the dunce of the day. The year before in the mornings at the old bus stop there were girls who would cry or become physically ill when the school bus came into sight. My own sister would sometimes become sick to her stomach as the bus pulled up to the curb and have to stay home for the day.

My situation was made more difficult by the fact that I was so very much smaller than anyone else, which apparently among the other larger children who had been forcefully schooled in the ways of Christ was seen as an open invitation to violence. At Blessed Sacrament we all took part in the initial testing for the Salk polio vaccine. Most of us knew somebody who had problems from polio. In the old neighborhood my sister and I sometimes played with a little girl who had been crippled by the disease and who wore braces on her legs and walked with crutches. At school we were all used as guinea pigs in two separate vaccine experiments, one of which was administered orally and the other by injection. Each trial included three doses over time, I believe. We were all given physicals in the gym beneath the school before the experiments began. I remember being mocked and teased by some of the other boys about my weight. "Tin can thirty five!" is what Jamie Collins called me when he found out that was my weight.

From our new home, even though the bus stop was one block from the old one, we rode a different route with different kids and the three kids who got on and off with me at my stop began to push me around every morning and to beat me thoroughly and mercilessly every evening as I tried to get home.

Fred Podanski was the blonde, crew cut son of a market owner downtown. Fred lifted weights and was the biggest and most muscular of the three. Kevin Dugan was the fourth of nine children of an Irish civil servant. Allan Bickard was the oldest of six children of an office type.

From the first day I got off the bus until the last day I attended Catholic school I don't remember a single day when I was allowed to go home without being punched, thrown, twisted, and bent by the three bullies who were each a head taller than me and in the beginning, each outweighed me by double. The humiliating beatings and other torment often went on each day for half an hour I am sure.

I would sometimes try to sit near the front of the bus so I could run off first and try to beat them down the street. I soon found out that was hopeless. They were all faster than me. I could barely lug the heavy leather book bag full of textbooks anyway and had to bend myself to one side just to walk with it. I also had to carry a metal lunchbox with a thermos inside and I remember on more than one occasion the bullies wresting it away from me and then flinging it on to the sidewalk breaking the glass liner in the thermos.

I remember Podanski's voice behind me as I struggled in pain to lug the briefcase down the street thinking I might have gotten a step on them. "Where do you think you're going, Tommy?" he taunted as they easily caught me just one house away from the bus stop, pulling my books and lunch box from me and dragging me back to the side yard of the corner house by the bus stop where the years of daily beatings took place.

Sometimes they would pretend to be teaching each other wrestling holds or combinations of punches using me to demonstrate.

"Here, see how much more it hurts when you punch down into someone's stomach instead of up into it."

"I don't believe it. Let me try."

Sometimes they would hit me in turn and sometimes they would pile on all at once. I remember wincing and groaning in pain but I do not remember crying, ever.

I was punched in the face, flipped over, thrown and knocked to the ground, had my arm twisted, fingers bent back and was pulled by the hair and even swung by my ankles and slammed into a tree. I was picked up over their heads and thrown into thorn bushes, my new front teeth were chipped by having my face pushed into a granite wall and one time someone tried to force me to eat dog feces.

They would take my catechism from my book bag and threaten to damage it or keep it knowing that I would be frightened of retribution by the nuns. I tried for years to fight back with enough force to hurt them and back them off. I tried to get them to fight me one at a time. The one time I was able to convince them to let me fight Bickard alone I got four or five good punches in and had him looking worried before one of the others tripped me and Bickard dove on top followed by the other two. I tried riding the several miles to school on my bicycle to avoid them, pedaling frantically up and down hills to try and beat the bus home. When I did avoid them they would track me down later in the day with increased malice.

Podanski was the least bright of the bunch and was out to show he was tough. Years later he applied to be a State Policeman and asked another neighbor's father, who was a government big wig, to write him a reference letter. The neighbor's son had also been bullied by Podanski and the letter thankfully put an end to Fred's hope to become a gun toting bully on the state payroll. Dugan was the least enthusiastic of the bullies and on a couple of occasions actually called an end to the sessions when he could see how badly I had been hurt. Coming from a home where he was the fourth oldest of eight boys and one girl he probably experienced some sympathy for being picked on. Allan Bickard was probably the brightest of the three but far and away the most vicious and relentless. He had the same shocking look of deliberateness and cruelty that I had seen on the mean little 'lectricity kid back in in Seattle. I can remember Bickard continuing to rain punches on me when I was flat on the ground and too weak and hurt to do anything but try to cover my face. I don't know if any of the three ever give a moment's pause to reflect on the damage they did to me with the hundreds of beatings I took, but I still think of them and especially of Allan Bickard. I have sometimes thought about what would happen if I ran into Allan. I pray it doesn't happen and I think he should pray for that as well, because as long ago as it was and as much work as I have done to heal myself from all the hurt I experienced I don't know if I could control the rage I feel when I remember his vicious sneer and squinting eyes.

I remember finally making it to the front door of the house on Latona Avenue one day. My nose was still bleeding, my clothes torn and my arm was inflamed from lugging the heavy book bag. I was exhausted as I reached the front door where my mother met me. I looked up wearily and spent. "Oh, God," she yelled as she looked up and away from me. "Not another pair of pants!" She turned on her heel and strode back into the kitchen, keeping her back to me as I climbed the stairs and disappeared into my room.

I had frequent nose bleeds during this period. I began to wet my bed nearly every night and I developed several pronounced tics or spasms. I would pop my thumbs at the base until I could actually hear one of them squeak. I began to twitch my nose, scrunch up my stomach and chest and developed a twitch or tic in my neck that made the tendons stand out. My parents occasionally took me to the doctor to have the veins in my nose cauterized, believing I had some kind of congenital disposition to nose bleeds. It became evident later in life that my nose had been broken several times as a child but no one was apparently attuned to it at the time. I also have compression fractures in several vertebrae as well as shoulder, knee, neck and elbow problems that have no other explanation.

One night some of my sister's girlfriends from across the street came to my house to announce that Richard Sifford was coming down the street to beat me up. Richard was a grade behind me and went to a different school. He was however larger than any of my daily tormentors. My father was outside trimming bushes and stood by as a crowd of twenty or so kids gathered to watch. Richard showed up on his bike, got off and came up and punched me as hard as I had ever been punched. He drew blood with the first hit and as hard as I tried to hit back I was no match. My father's only attempt to intervene was to yell "Kick him in the nuts," at one point but I don't think I even knew what that meant. The entire front of my shirt was soaked with blood along with the waist of my jeans. I limped inside bewildered by why Richard Sifford, who I didn't even know, would want to fight me or hurt me. I didn't understand how my father could stand by without protecting me or at least stopping the fight when I was bleeding so badly and it was so obvious that I had no chance.

When Richard finished pounding me he merely got back on his bike and rode up the street. His father, Colonel Sifford, who was a big wig at a nearby army post, strolled down the street a minute or two later and struck up a conversation with my father. He explained that Richard had been bullied by the three kids who were after me all the time and that he had taught Richard how to box as a result. I felt feverish and I held a handkerchief to my still bleeding nose as I watched through the screen in my bedroom window while my father and Colonel Sifford talked together casually. My father never came in to check on me and instead walked off up the street with Colonel Sifford. Neither did my father teach me to box or even talk with me about what had happened. The beatings at the bus stop went on for four years longer.

CHAPTER 8
Vegetables

"Did you eat your lima beans yet?" My father asked threateningly. He lowered the newspaper so he could glare over it at me. He had pulled an arm chair out into the center of the living room and rotated it so he could sit facing me as I sat alone at the dining room table.

"No and I'm not going to," I answered back, trying to match the heavily moderated malevolence in my father's voice.

There were only two vegetables I had trouble getting down. Lima beans and peas both made me gag. I had no trouble with wax beans, green beans, broccoli, cauliflower, carrots, beets, asparagus or even spinach and Brussels sprouts. No matter, according to my father, vegetables were good for us and we had to eat every last pea or bean that was put on our plate.

At one time I had tried slipping a lima bean to our dog, Andy. Right away he started tapping his tongue about and scraping it against his teeth like he had a hair in his mouth, finally lowering his head and dropping the bean onto the carpet with a final cough of relief.

My sister's problems were worse. She could only eat wax beans and green beans without real trouble and everything else was a problem. The vegetables we were served were not fresh or in most cases even frozen but rather canned, so asparagus for example was not crisp and mild but salty, soggy and yellow. At age sixty-five she still has an aversion to most vegetables and her eyes tear up as she remembers hearing her friends from the neighborhood playing happily outside while she was forced to sit for literally hours, alone at the dining room table, staring at a plate of uneaten Brussels sprouts or stewed tomatoes. Over time she developed a few of her own unique strategies to deal with the issue. Our mixed breed Miniature Collie, Andy, would help her out with certain of the vegetables. Other times Barbara would slip lima beans or broccoli into her napkin a few at a time, then excuse herself to the bathroom where she would flush the slimy overcooked veggies down the toilet. At one time she had been caught sneaking the goods into a wastebasket and had subsequently learned to get rid of the evidence.

I would sometimes manage to avoid the otherwise inevitable confrontation with my father by swallowing peas whole with swigs of milk as if they were pills. I disliked milk as well so this was no easy task. Lima's were another story. They were too big to swallow whole and the mushy waxy texture was just too much for me.

My mother would cajole us about starving people in China that would be grateful to have lima beans to eat. Barbara and I would of course offer to send them a year's supply of anything they would take. My father's absolute insistence that the vegetables we were served had to be eaten because they were good for us rang hollow to me. He himself would pick out and push aside green bell peppers from a casserole we were served occasionally.

I knew full well there were other vegetables and plenty of other items that could have been substituted that would have been equally nutritious while having a lower gag factor than the ones we had trouble with. By dinner time I had already taken a beating trying to make it home from school and had often been scolded and sometimes pinched, pushed or whacked by my mother between then and dinner time. Parents forcing us to do something that made our stomachs turn was just more of the same.

Dinner time was much later at our house than at the other kids homes in the neighborhood, as Dad usually didn't get home from work until close to seven p.m. We seldom saw my father in the mornings. Most weekdays he would leave for work before we came downstairs for breakfast. Sunday my mother, sister and I would go off to Catholic mass and my father would go to a Methodist church service. Saturdays Dad would usually go into work for a half day. When he didn't work he would disappear to the barber shop or gas station or on a rare occasion to the hardware store, coming home with a furnace filter, a faucet washer or toilet tank repair item which would keep him fully engaged for much of the rest of the day. My father was admittedly mechanically inept, in spite of the fact that he ended up as CEO and Chairman of a large corporation that manufactured sophisticated pumps and other oil moving equipment, diesel engines and even steam turbines for nuclear submarines. A simple household repair undertaken by our father could easily result in a house full of smoke, dust or ankle deep water. Mumbled curse words, skinned knuckles and second and third trips to the hardware store to trade in the wrong parts or to replace ones that had broken trying to make them fit, were all part of a normal Saturday home from work.

Whether it was a weekday, a Saturday or a Sunday the first order of business before dinner was cocktails. Old Fashioneds were my parents' drink of choice when I was in grammar school. In later years my mother switched to top shelf scotch on the rocks and my father to V.O. blended liquor. In any case the drinks were always doubles and sometimes they had three rather than two before dinner. Barbara remembers them having four drinks at times. It of course never occurred to me that this was anything but normal. Looking back, I now realize that pretty much every day of my childhood I sat down to dinner with parents who had each just consumed four to six or even eight ounces of hard liquor!

The sound of cracking ice cubes was a sure sign the old man had made it home from work. He had a device made specially made for the purpose. It had a kind of half round metal ball attached to a handle by a long thin piece of spring steel. My father would adeptly vibrate it with his wrist motion, cracking the ice into small pieces and allowing them to slide from his other hand into the glasses. The particular alcoholic potions would then be carefully and methodically measured and poured into precious painted and engraved Old Fashioned glasses along with the bitters and other ingredients. Finally, the drinks would be garnished with a carefully and exactly cut orange slice and a cherry. The second round would come together much more quickly and with less flair and finesse. My sister and I were both veritably starved by this time but still would have to endure the time it took the parents to finish the first drink, pour the second one and finish at least half of that before we would be called to the dining room to eat. The other kids in the neighborhood had finished dinner by five thirty or six and in the warmer months had been out playing in the lot across the street for an hour and a half or two when we sat down.

By the time dinner was actually served my stomach was quivering with hunger. If I looked down to see that peas or limas were the vegetable of the day my appetite quickly failed and was slowly replaced by a Gordian knot in my chest. While I stared down at the soggy unappetizing greenish yellow chokers, my father ate like a hog at a trough. He bent low over his plate and scooped food into his mouth at a rate that could not possibly have allowed his taste buds to register any but the grossest, most obvious of flavors. He slurped and chomped and was always finished eating long before anyone else at the table was a third of the way through the meal. I would sometimes look to my left at my mother, hoping she would say something to my father about the noises he was making. I wondered how my father was perceived at the numerous business dinners and lunches that he attended. At even a very young age I was embarrassed for him.

My Dad used to brag about getting second helpings at mealtimes in the children's home his father had placed him in Amsterdam, New York. His mother had died two weeks after he was born. In those days (1920) the medical doctors had recently taken over the supervision of childbirth from the midwives who had handled the chore for many years. The midwives had considered childbirth as a natural process often using time tested herbs, teas and other simple methods to aid and comfort the mother in delivering the baby. When the doctors took over the field, their training and belief in themselves as warriors against sickness and disease naturally colored the way they saw childbirth. To them a pregnant woman was essentially "sick" and had to be treated. They the doctors "delivered" the baby, and the woman who had undergone this ordeal had to be treated for a time afterwards. The fact that there were women prior to this that had delivered babies during the harvest and gone back to working in the fields that same day was seen as some kind of folklore by the doctors who had gained a new and lucrative area of practice when they took over from the midwives, eventually moving to outlaw them. As a result, a doctor insisted that my father's mother stay in bed for fourteen days after the birth of my father. During that time, she developed Phlebitis, known at the time as "milk leg," from her lack of activity… and she died.

My grandfather, Thomas Wheaton Bixby who was a foreman at a glove factory in Gloversville, NY, placed my father with my father's aunt and her husband. By the age of four they claimed my father was unmanageable, often screaming and crying in the night. By then Grandfather Bixby had taken up with my step grandmother, Harriet, who he would live with for twenty-five years before marrying her for another twenty-five. That was actually his third marriage, his second being Talitha Raison, my dad's mother. His first wife, Bertha, ran off with the milkman who happened to be married to Grandfather Bixby's younger sister. When my father was four, my grandfather committed him to the children's home, later claiming that his girlfriend Harriet was too mean and he thought it better to put his son in an orphanage than subject him to her.

My father boasted about never missing seconds at mealtime in the children's home. He told us that there was enough food for only the first two or three done eating to get second helpings. My father said he never missed seconds even once in all the years he lived in the orphanage.

At our home on Latona Ave., after wolfing down his dinner my father would often retrieve a newspaper from the living room and disappear into the small bathroom that was far too short a distance from the dining room. The sounds of paper rattling as the newspaper pages turned could be easily heard from the dinner table along with grunting and other sounds that did nothing for the appetite.

"Eat your vegetables," he called out loudly as came back to the table, and slapped open the newspaper. If ice cream or something else was being served for dessert the parents would go ahead and eat their portions, making it clear that my sister and I would not get any until we finished the vegetables. When my mother would retire to the kitchen my father would sit at the head of the table reminding us every few minutes that his patience was growing ever shorter.

To me it became yet another confrontation with a bully who had no justification or right to treat me that way. I had learned to steel myself against just such threats and after years of frustration trying to gag down the mushy overcooked beans, I finally just refused to go along.

My father, after sitting at the table with me for well over an hour without any progress had stormed off. Halfway through the living room he had second thoughts and trying to make it appear as if it had been his original intent, he stepped back a bit and then to the side, pulling an armchair into the middle of the room. He spun it around, plopped into it and rattled the newspaper as he blew out a breath of frustration. After seething inside for a minute or two he slapped the paper down into his lap and blurted out loudly, "You're going to sit there until you eat those vegetables. If you don't eat them tonight you're going to get them cold for breakfast."

I stared back at him eye to eye without answering. After a minute or so he rattled the paper and raised it up again. From behind it came, "You'll eat them if I have to sit here all night." There were several more similar exchanges over the next few hours until finally a little after eleven o'clock, without a word my father jumped out of his chair, strode across and into the dining room, leaned across the table and slapped me on the ear with the heel of his hand. I steeled myself and simply stared up at him defiantly. Still holding the now partly crumpled newspaper with one hand he roughly snatched the dinner plate of limas and stomped wildly toward the kitchen. As he reached the threshold he flung the plate across the room smashing it against the sink sending shattered porcelain and soggy legumes in several directions. His last words to me that night were screamed. "Go to bed goddammit!"

For my father I am sure it was about control. If he said it was going to be this way, then by God there was no other way. Everything in the house ran quietly and smoothly according to the rules, which were added to at least weekly, or so it seemed. Usually a vicious look and a stern scolding from my mother or a simple quick side-glance from my father was enough to keep us toeing the line. Any indication or resistance from me could cause my father to fly into a short lived but violent rage. "Get his pants down, Mary. I'll get the belt!" I can remember trying to dive under a bed to escape and my mother pulling me out by my ankles while my father whipped me with his doubled over leather belt.

Mr. Jim Gutzwiller, one of my father's co-workers, stopped at our house to pick up my father. I thought it was a funny name and as they got into the car in our driveway, I waved goodbye. "Goodbye Mr. Guts- willy- willer," I called out. My father ran back up to the house and hit me so hard in the face with the heel of his hand that he knocked me completely off my feet and a good three or four feet through the air. He never said a word and I was still lying on the floor seeing stars as he turned his back and walked back out of the house.

My sister and I started to giggle over something at supper one night. We were sternly warned to stop which only fueled the laughter even more. "Go to your rooms," my father yelled, "and just wait there until we decide how to punish you." We waited a half an hour or more until my parents calmly climbed the stairs, came into my room and my mother held me down while my father beat me with a belt. They then went down the hall and hit my sister with the belt. To this day neither my sister or I can remember a single time when I cried as a child and my sister says as horrible as it was to be hit it was worse for her hear me being hit while she waited for her turn.

CHAPTER 9
Side Effects

Fear or more accurately nearly infinite terror was the predominant feeling during my initial breakdown in the summer of 1970 at the age of twenty-two. When the photosensitivity from the Thorazine side effects caused my skin to burn, that fear was instantly activated and energized. My mind raced to create an explanation that would explain it and I became even more convinced that I was not only dead but also burning in hell. My fear was extremely powerful in a way that before the breakdown, I had only experienced in nightmares. In a nightmare, fear that strong usually wakes us up. This was fear that was as blindingly strong but it was not possible to escape. Think of fear as a liquid. Imagine that I felt like I was suddenly dropped into a swimming pool full of it at midnight and held under until the fear would eat away every molecule of my being and I would dissolve into and merge with the fear; my entire consciousness, my very existence becoming pure and unadulterated liquid fear. It was really that strong.

In those moments when I was not in active terror, I was in a state of anxious hypervigilance, much of my attention focused at the edge of my consciousness expecting the fear to resurface at any time. The rest of my attention was directed outward, also in a heightened state trying to micromanage my life in a way to avoid anything and everything that could trigger another terror attack. It was devastating and it was exhausting.

When Dr. Silverston casually informed me that the drug he had given me caused the side effects and effectively the further breakdown of my mind as I feverishly sought an explanation for the symptoms, I quickly ignored the flash of anger that ran through me. I was now even more desperate for an answer that would fix my problems. I wasn't about to alienate the doctor. Although putting any faith in him seemed very faint hope, it was the only hope I had.

He quickly wrote out a new prescription for another drug at a much higher dose. The drug was Mellaril, generically Thioridazine, another phenothiazine derivative that blocks nerve impulses. I believe the dosage was 200 mg four times daily, an extremely high level. Again no explanation was given to me about why I was given it. Dr. Silverston still held to the idea that my problems were from alcoholism and he offered no indication as to how the drug might work or what side effects I might anticipate except for dryness in the mouth. I was simply told I would feel better on this drug than the other one.

Side effects from Mellaril can include tachycardia, hypotension, dizziness, fainting, dry mouth, oral moniliasis (fungal infection), drowsiness, blurred vision, toxic retinopathy and blindness, discoloration of the skin, contagious rashes, bladder paralysis, edema, constipation, diarrhea, vomiting, impotence, ejaculation inhibition, feminization, paralytic ileus or obstruction of the intestine, declining white blood cell count, withdrawal symptoms, Parkinson like symptoms, akasthisia-which includes the inability to remain still, inability to concentrate, pacing, chewing and lip movements, finger and leg movements; convulsive seizures, depression, feelings of unreality, neck, eye and back spasms, stereotyped involuntary rhythmic movements of the face, mouth and extremities, involuntary gross motor "jerking" of the extremities, face, jaw and tongue; photosensitivity, weight gain, diabetes like symptoms from hyperglycemia, whitish brown granular deposits in the cornea and lens of the eye, hyperthermia and hypothermia, and toxic psychoses including visual hallucinations. Many of the mannerisms and stereotypes we attribute to disturbed patients are actually caused by the so called cure, not the disease.

Although I had a number of these side effects, I suffered most in silence believing they were part of my general deterioration. By the end of the first day on the drug I had lost almost all feeling in my limbs and even in my face and hands. While the drug had reduced the level of fear, it had severely numbed my senses and shut down my thoughts almost completely. I could barely taste or smell food and colors all took on a grayish brown cast.

That night I lay down naked on the cold tile of the bathroom floor trying to feel any kind of sensation in the skin of my back. I felt as if I existed inside a gray smoke stained Plexiglas box and my body and everything outside it were far removed from me. Even the icy tile barely registered against my skin. The drug had dulled me beyond the point of numbness. I had lost nearly all feeling in my body. For months I had struggled with fear that my body was dissolving into space, that my soul would be exposed to the spiritual energies of the universe, the evil thoughts and demon avatars that roamed the astral regions. This was different. I did not feel as if I was leaving my body nor that I was about to become more sensitive to the energies that pursued me. This time I felt trapped in a body that was essentially dead, unable to experience anything.

Throughout my disease, in spite of many hours, days and even months of suicidal despair and terror of the infinite, some small and deeply hidden part of me refused to let go entirely and give in to the horrible idea that I was damned for all time and beyond. The terrible memories of beatings I had suffered through as a child were often with me as were memories of the prison riot at Fort Dix, some very particular memories of combat in Viet Nam, of nearly dying from the malaria symptoms and when I was swept under in the river at Bong Son trying to retrieve the dead body of a fellow soldier. In all those situations there had come a moment when I had felt lost, a moment when it felt that there was no point in fighting on… that essentially I was as good as dead already. In every case I ignored those feelings and somehow struggled on. Something had caused me to fight on when there was nothing left to fight with, to hang on. Studies have shown that men who perform best in overwhelming combat situations were those who were forced to fight a lot as children. As hopeless as my situation assuredly was, something in me refused to accept it entirely. My sense of right and wrong was so deep that I could not quite believe that what was happening to me was possible. If things were what they seemed, it made me so angry that I was determined to fight on if only to give God one good black eye before he sent me to hell.

Desperate to feel anything at all I began to masturbate as I lay with my bare back against the tile. There was nothing remotely erotic or sexual about it. I was simply trying to feel anything at all. When I finally reached climax it was barely perceptible. Suddenly instead of any release I was seized by a painful muscle spasm in my groin which somehow pinched off and blocked any ejaculation. I did not recognize this as a side effect of the Mellaril, which it was. My mind was too dulled from the drug to sort out or understand what had happened. I simply added it to the list of things that were "wrong" with me. My mind returned to obsessing over my ultimate eternal fate. My body was so dead to me while on the high levels of the anti psychotic drug that I began to have symptoms not unlike those experienced by subjects in sensory deprivation experiments.

I became afraid to open the door leading from the bathroom. I became convinced that the world outside had disappeared. The only thing that existed was my now nearly dead body and the walls of the bathroom which were nothing more than an illusion. Beyond the door was an eternal black void or for all I knew, fiery demons waiting to devour me. I stayed low on the floor, too anxious to even stand and look out the window. I stayed there for perhaps two hours before the obsession subsided enough to allow me to risk opening the door and crawling the few steps to the safety of my bed.

In the morning Tom Bird was at the door at 7 A.M. He was a high school friend that I had recently reunited with when he asked me to be in his wedding party. Tom had entered college immediately and gotten a civil service job with the State of New Jersey on graduation. He had just married a seventeen-year-old from his neighborhood, Linda Culwick. Tom had offered to introduce me to golf and we had apparently agreed to go out to a municipal course on that cool crisp Sunday morning. I had forgotten but stumbled into my clothes and borrowed my father's clubs from the garage. There was some steam rising from the sewer drains at the curb as we left the neighborhood and I pointed out quite casually to Tom that the swirling white cloud had materialized into a small knight in armor riding a likewise miniaturized horse. The knight was wielding a lance and was charging hard at the car as we drove by. I didn't look back at Tom to see his reaction as I was fascinated by the image, and was following its movements carefully as we pulled away.

As we turned up Route One near Lawrenceville, I saw Lucille Ball hitchhiking alongside the road in front of the Sleepy Hollow Motel. We passed her by. "That was Lucille Ball," I said. "Don't you think we should give her a ride?"

"That was not Lucille Ball!" My friend said alarmingly. He stared at me for several moments. "I know what you guys are doing!" he said accusingly. He was referring to me and the other two high school friends of the foursome that had re-met at Tom's wedding. One of that group, Steve Smith, had become a heavy drug user while in the military stationed in Japan. Tom, I believe, was convinced that I was high on LSD which was making its rounds that year.

I strained around in the seat to look back at the hitchhiking woman.

"That's definitely her, Tom," I quietly insisted. "She's got the bright red hair and everything."

He stared at me again for a long moment. "I'm taking you home," He said and he made a U-turn at the next jug handle, drove me straight back to my parents' house and dropped me off without another word. That was in the summer of 1970. The next time I saw Tom Bird was in 1989. I also missed my chance to try golf that morning. It was twenty-two years later before I finally got the chance again.

That night as I sat on the edge of my bed talking with Dot Franks, the wife of my race car partner, she nodded sympathetically. When her husband Ed and I had become race car partners Dot and I had struck up an immediate friendship. I had spent hours in her kitchen drinking coffee and talking into the night with Dot and her friends. The night I left for Vietnam Dot kissed me goodbye in a way that made me feel that she cared very much about me. While I was away she sent me notes and cards, one or two that were a just a little bit on the flirtatious side. Both of us knew that nothing would ever happen between us. She was married and four or so years older than me. I was far too moral and far too anxious to have initiated anything myself, and Dot had four kids and a husband to keep her in line. Nonetheless we were two souls who seemed to understand each other better than most and had sometimes confided in each other when we needed to.

The first time I saw Dot after the war I sat across from her at her kitchen table on Prospect Avenue. She told me that a while back Ed had suspected her of having an affair and she said she was afraid he might harm himself if he caught her cheating on him. I was not about to disrespect my friendship with my partner anyway but that instantly and forever ended any boyish fantasies I may have had about dallying with Dot. It also only strengthened our friendship and understanding of each other.

That night as I shared with Dot some of the fears I was living with, I heard my mother and father walking up and down in the hallway outside my bedroom. As I looked up toward the door, my mother asked quite anxiously, "Tom, who are you talking to?"

"Dot's here," I answered casually.

"Tom," my father said rather sternly, "open the door."

I got up from the edge of the bed and crossed the couple of steps to the door. My parents were standing together shoulder to shoulder leaning forward as if they had been listening at the door.

"Tom, Dottie is not here." My mother was alarmingly adamant, but I was well used to my mother raising her voice to make her point her way.

I shrugged dismissively and turned back toward the bed to indicate Dot ...but she was gone! "Where the hell has she gone?" I thought. I quickly surveyed the tiny room and not finding any sign of her I dropped to the floor and peered under the bed. I ran to the closet and not finding her there I brushed past my parents in the doorway and quickly dragged a chair from the room at the end of the hall into my bedroom. I pushed the chair into my clothes closet, climbed up on it and slid back the trap door in the ceiling. I pulled myself up into the attic crawl space, the only other place I could imagine Dot could have disappeared to.

At that point I heard my parents retreat into their bedroom and close the door behind them. They talked in hushed tones then dialed the phone and talked on it for some time.

In the morning I was taken to the emergency room at Mercer Hospital in Trenton and admitted as a patient of Doctor Jack Ward. My parents had tried to reach my psychiatrist, Dr. Silverston the night before. His answering service had called back and said that Doctor Silverston would not be able to see me anymore because he was going on vacation soon. My parents were referred to the office of Dr. Ward and his receptionist had called in the morning with instructions to bring me to the emergency room. She said Doctor Ward specialized in people who had problems such as mine. I was admitted to the hospital suffering from an Acute Schizophrenic Reaction.

PHOTO BY DARRYL SHUMAKER

CHAPTER 10
Dr. Jack Ward

I sat alone in a wheelchair in the emergency room hallway. Just twenty-two and a half years before when I had been born there, Mercer Hospital was surrounded by one of the most beautiful and desirable neighborhoods in the area. Now both the old brick front hospital and the neighborhood surrounding it were deteriorating quickly. Entering through even the newer yet poorly maintained wing which housed the ER, only added to my feeling of impending doom rather than giving me a sense that some kind of healing was about to take place.

Unfortunately, my birth wasn't the only other time I had been in Mercer Hospital as a patient. It was also far from the only hospital I had been in during those first twenty-two years. I had been rushed to Oakland Children's Hospital in California suffering from convulsions and near brain meningitis brought on by an extreme case of the mumps. A drug that was given to me for the high fevers stopped my heart and I had been taken by ambulance in the middle of the night and kept alone in a quarantine room for several days. I had been more than a week in Mercer when I was about eight for a tonsillectomy operation with complications. In junior high I was in Mercer Hospital again for an operation, under general anesthesia, on my hand which had been stomped on by a bruiser from the "special class."

In Vietnam I spent six weeks in 67th Evacuation Hospital in Quinhon, two weeks on each of three separate occasions, for malaria and gastroenteritis symptoms. I was treated for minor shrapnel wounds to my leg in the 173rd Airborne Brigade's forward aid station. I had also been in the emergency room where I now sat at Mercer at least a half dozen other times for other injuries to my hands, leg and head as a child.

"At least I won't have to get stitches this time," I tried to cheer myself up but that thought was quickly replaced by one of those awful internal voices whispering, "unless they give you a lobotomy!"

My mind and body were so numbed by what was essentially an overdose of the drug Mellaril as prescribed by Dr. Silverston that even the thought that I could be facing something as extreme as a lobotomy barely registered any anxiety in me. I really had no idea what might happen to me although I had heard people talk of a mental ward on the top floor and I knew about things like electroconvulsive shock therapy. Although those kinds of things were worrisome, the idea that I had a disease that was sometimes treatable was still less fear provoking than the notion that I was dead and living in some ever changing and eternal hell realm. I was so numbed and my mind so deadened by the drug that even those thoughts came and went infrequently and without the usually associated panic attacks.

My internal clock was as inoperative as the rest of my senses. It may have been three hours or it may have been only thirty minutes that I sat alone in the hallway of the emergency room. A hundred or more people may have passed by or only twenty. I had expected to see the new doctor but as far as I know he never came. Instead I was eventually wheeled off to different offices and laboratories within the hospital. I was given an EEG or Electro Encephalogram to check my basic brain function. I had blood drawn on several occasions over several hours and in between was made to drink a foul tasting milk shake sized glucose solution.

When I was finally deposited in a semi private room, I was surprised to learn it was not on the mental ward but an ordinary inpatient floor. Between the effects of the Mellaril and of glucose solution I was quite foggy when I reached the room but I remember my father talking not to me but to the middle aged man in the next bed who was recovering from some minor surgery. My father was complaining to the man about the length of my hair. My hair nearly reached the tops of my shoulders as was fashionable in those years. I let it grow after separating from the Army, more as an experiment than as any political statement or as a symbol of any group.

My father sat not at my bedside but at the bed of the other patient and complained bitterly about my hair and the assault on his values and his way of life that it represented. The other man listened attentively before dismissing my father's long and agitated monologue with a simple sentence.

"Jesus had long hair," he said.

That night or perhaps the next morning, Dr. Jack Ward, my new psychiatrist, made his first appearance. He sat down next to the bed and talked with me at length. He told me the visual and audio hallucinations were surely a reaction to the medicine Mellaril. He would immediately reduce the dose to a much lower level. He also told me that many of my thought problems and terror attacks were the result of chemical imbalances in the brain. He drew a chart which demonstrated the way normal people metabolize sugars and then superimposed a graph line which showed my own sugar levels over time as had been recorded during the Glucose Tolerance Test he had ordered earlier.

He went on to explain how my heavy drinking to suppress the schizophrenic symptoms had actually exacerbated them. The symptoms appeared in part because of low blood sugar levels to my brain. As borne out by the Glucose Tolerance Test, the more sugar that was introduced into my system, the more my system would overreact with insulin, burning the sugars too fast and leaving me with extremely low blood sugar levels. The alcohol was metabolized as sugar and had caused a cycle which continuously led me back to the complex of symptoms caused by low blood sugar to the brain. He told me it was a permanent condition. He called it hyper-insulinism, alternatively hypoglycemia. He said it could be controlled with very high protein diet and with a regimen of very high doses of vitamins and other medicines. He placed me on what he called a Salzaar diet, which called for one hundred grams of protein a day, very little carbohydrates and no sugars. He advised me to stop drinking entirely as the alcohol was metabolized as sugar and he told me to avoid coffee and other stimulants. In addition to drastically reducing the levels of the anti psychotic drug he also placed me on a "Megavitamin" regimen which included six thousand milligrams of Niacin or Vitamin B-3, three thousand milligrams of Vitamin C, three thousand I.U.'s of Vitamin E and three hundred milligrams of Vitamin B-6. In addition, he placed me on a medicine called Deaner and another called DBI, which was an anti-diabetic drug that would help control the insulin spills from my overactive pancreas.

This was certainly what I had hoped for. The Doctor's "Orthomolecular" approach to my disease explained it in a logical, physically based way. It was something of a comfort to hear Doctor Ward speak to me about it in a matter of fact down to earth way. I was more than willing to go along with the diet and anything else that could help.

The visual and audio hallucinations disappeared immediately when the massive doses of Mellaril were reduced significantly. I stayed in the hospital for a full two weeks. My energy level was very low initially but my anxiety level was also reduced. The dietician made two visits to my room to be certain I was able to eat enough of the relatively bland diet to keep me nourished. I was still quite thin. I was six feet tall and weighed only a hundred twenty pounds. I had returned from Vietnam weighing only one hundred twelve as a result of the malaria/ gastroenteritis hospitalizations and it would be another seventeen years when I finally topped one hundred thirty-five as I turned forty years of age.

I learned that Dr. Ward's approach to schizophrenia was based on research by two famous Canadian doctors, Abram Hoffer and Humphrey Osmond, who had named their theory Orthomolecular Psychiatry. They essentially rejected the "psychiatric model" of schizophrenia, creating instead a "medical model" based on the work of Dr. Linus Pauling who had coined the term Orthomolecular Medicine. Pauling wrote that he advocated "the preservation of good health and the prevention of disease by varying the concentration in the human body of the molecules of substances that are normally present, many of them required for life, such as vitamins, essential amino acids, essential fats, and minerals."

The Orthomolecular Psychiatrists believed that the lack of nutrients or the presence of toxic materials in the body can cause "Dysperceptions" in susceptible individuals. Under this belief system the schizophrenic may experience distortions in the size or color of objects, time may seem speeded up or slowed down, spatial relationships may be distorted and even the perception of one's own body may be terrifyingly inaccurate.

Ordinary incidents may seem bizarre and other people's behavior can appear weird or even shocking. Imagine simply that a person in a normal conversation suddenly seems to the psychotic to be much closer than is comfortable. He seems larger than life and is talking much faster and much louder than normal. This can be bewildering and frightening to the schizophrenic whose mind under stress may reach for an explanation to justify these distortions. That explanation may make sense only to him much in the way things which are illogical to us in our normal waking state make sense when we are dreaming.

The doctors believed that while psychotic disorders can be triggered by trauma, shock or long term stress, that the main factor was a biochemical one. Hoffer and Osmond actually injected themselves with adrenochrome, a substance that earlier researchers found present in urine and blood samples from schizophrenics. Dr. Osmond found his perceptions immediately altered. Colors and lights appeared differently. His physical surroundings seemed "sinister and unfriendly" and he found himself feeling indifferent and even annoyed at other people. He even found himself disturbed by what he believed were "covert glances of a sinister looking man" in a restaurant.

Hoffer and Osmond had also conducted experiments on the effects of hallucinogens like mescaline and LSD and had eventually discovered that Vitamin B-3 or Niacin helped to alleviate or reduce the effects of those drugs. They had gone on to develop a regimen of what they called Megavitamins which where helpful in mitigating the symptoms of schizophrenics. Dr Ward started me on Niacin and then switched me to a similar compound called Niacinamide when I developed a skin flush. The difference in my reaction to the Niacin skin flush compared to the meltdown I had experienced with the photosensitivity from Thorazine was significant. With the Niacin, Dr. Ward had warned me ahead of time about the possible side effects. Dr. Silverston gave me no warnings and my mind had raced out of control imagining all kinds of horrors to explain the extreme burning in my body.

My main memory of that first of my three stays in the hospital is of being tired. Both my thoughts and my physical sensations were still quite dull from what was essentially an overdose of the anti psychotic medicine given by the first psychiatrist. The most severe of the panic or terror attacks subsided. My mind still raced on about what eternal horror I might be trapped in but now I had an alternative way to look at what was happening to me. The idea that my terrifying experiences might be physically based helped me to balance my other thoughts. Even the mere suspicion that I was suffering from something that could have a physical cause helped me to want to believe again that there was in fact a concrete physical or material reality, something which I had doubted for many months. For some time, my perceptions had been distorted, my sense of reality and even time so fluid that an eternal realm of mind had seemed more credible than the solid and predictable world that most people accept without question. In the hospital, time slowed back down. Eternity was too far away to be concerned about. Under the spell of my terror, reality at times had been no more than a temporary collection of fleeting fluid mental images that had no more solidity than the images in a dream. This had clearly been compounded when the massive doses of Mellaril had been introduced into my system dulling my senses to the point where I had to doubt whether even my body really existed.

The sheets on the hospital bed were stiff and scratchy but there was comfort in being able to feel them at all. I felt sick but I took some comfort in having a sick body that felt real. It was so much better than feeling as if I could dissolve into space or into some dream or alternate reality at any moment. The air or the energy in the hospital hallway felt thick. The movements of the nurses and technicians seemed slower, less urgent than the frantic world of my panicked mind for the last several months. I was uncomfortable in the hospital but that discomfort made reality feel more real… and that was a comfort in itself. That tiny and elusive sense of peace, that was really based on a feeling of sickness and discomfort in my body, was all I that had to build on. It was the basis for a small hope that I wasn't in hell.

CHAPTER 11

Office Visits

Dr. Ward's office was on a high floor in the Carteret Arms, a modern high rise apartment building. It overlooked the Delaware River at the edge of downtown Trenton, just a block and a half from the state capitol building. It towered many stories above anything nearby and may have been the tallest building in Trenton. Trenton had been a booming manufacturing town for many years but by 1970 when I was released from the hospital, the trend had begun to reverse significantly. For that reason, The Carteret Arms was among the last buildings to be built in the area. The motto "Trenton Makes The World Takes" had for many years proudly been displayed in giant lighted letters on a bridge that crossed the river to Morrisville, Pennsylvania. Now several of the letters were out and no one seemed to care enough to make repairs.

At its height Trenton had been quite a model city. Hill Refrigeration, Transamerica De Laval, General Motors, General Electric, Stangyl Pottery, Lennox China, Youngs Rubber, Congoleum, and Boehm Porcelain are just a few of the companies that had major manufacturing plants in Trenton. There were many smaller tool and die makers, electrical contractors and other businesses which were supported by the larger companies and also many building and road contractors which were kept in business by the trend in the fifties and sixties to move outward to the suburbs.

Trenton had a beautiful park with a small zoo, a working central shopping area downtown and a safe and comfortable bus system for getting around. In the fifties, my grandmother and her two sisters used to get dressed up, get on a bus wearing hats and gloves, ride down to the center of Trenton and see a movie then get on another bus and go all the way to the far side of town to eat dinner at a restaurant in the Italian section known as Chambersburg. On the way home, well after dark they, would change busses again downtown and arrive back home in the suburbs never having shown the slightest concern about their safety.

My grandmother died in 1957. By July of 1970, when I first saw Dr. Ward, Trenton had been transformed. Many of the original Trenton city residents had moved out to Ewing, Hamilton or other suburbs of the city. Only in the solidly Italian Polish section known as Chambersburg did the younger generation remain in their family homes or take up residence nearby. Many people moved into the Trenton from southern states or from the Caribbean seeking work in Trenton's many factories. Factory heads like my father were bewildered by the work ethic of many of these new employees. My father used to complain about the number of people that would drink on the job or fail to show up for work on Mondays without even calling in. As the quality of the workforce deteriorated many factories moved elsewhere, sent work overseas or simply shut down operations.

There were riots in Trenton in April of 1968 and most of the merchants who had been robbed and burned out left downtown for the suburban malls. The Mayor of Trenton, Arthur "Chirp" Holland moved into the center of the city to prove that people of different backgrounds could live side by side. After his home was broken into on multiple occasions and after both his wife and daughter had been attacked, he moved out of town as well.

By the time I had to go downtown to Dr. Ward's office, driving through Trenton had come to feel far less safe than when my grandmother took the bus or when I used to ride my bicycle through the park. Now as I drove past Cadwalder Park on the way to Dr. Ward's office I passed through an underpass beneath the canal that was built by my great grandfather, James Ross, an immigrant from Northern Ireland. The underpass was about a half block from Trenton's Junior High School #3. The little home I lived in until I was three years old was just four or so blocks up the river. When I was about ten I used to go to the indoor pool at Junior #3 during the summer to swim on some kind of a municipal program. The neighborhood, when I lived there in the late forties and early fifties was a quiet, safe family oriented neighborhood.

By 1970 the area was already quite run down and many Trentonians avoided it especially when school was letting out. The roadway narrowed at the underpass and mobs of kids coming home from Junior #3 would swarm out from behind the concrete columns of the underpass, jumping onto the hoods of cars, banging on the windshields and slapping the sides of cars as they went by. It happened to me twice, but looking back I think it was significant that I didn't feel any real fear when it happened. I had faced many bullies in my life. I had also run roads that were subject to sniper fire and ambushes almost daily as a Military Policeman in Vietnam. For whatever reason I didn't feel particularly threatened by the giant hordes of kids that tried to scare motorists as they drove through the area. I did find it depressing to see that my old neighborhood had changed so much that the people who grew up there were no longer comfortable or welcomed there.

In general, during the initial period when I was seeing Dr. Ward at his office, the extreme physical feelings of terror had subsided to an extent. My mind still raced anxiously with many of my obsessions about the nature of the universe and about my place in eternity. I could tie myself in knots in a matter of seconds if anything on the radio or in the newspaper or even a half a word overheard from a passerby triggered me. But as the extremes of fear slowed down or were held in check by medicines and diet, feelings of creeping despair replaced them. The drugs clearly dulled what I felt both physically and emotionally. My perceptions and sensations were not sharp. I was left with a feeling of "grayness." I felt no ambition. Plans did not come to me. I only cared about not feeling the terror. Everything in my life seemed dull, pointless.

A month before, riding up for so many seconds in an elevator would have certainly sent me into a panic about what supernatural horror I might find when the doors finally reopened. Now those same thoughts could find their way up out of the depths of my mind but I was too tired, drugged or numb to react very strongly.

Sitting in the waiting room at Dr. Ward's office I was surprised to see four or five other people there. At Dr. Silverston's office, I was always the only one there, either on the way in or out. At Ward's office there was none of the darkened tones, leather furniture or subdued lighting. The office was florescent lit, and the walls were sheetrock painted white. The carpets were industrial grade tan and the furniture made from simple cloth and chrome. It could have as easily been an insurance office as a doctor's.

Dr. Ward's inner office was tiny. There was barely room for his desk and the small simple office chair in which I sat. He had a matter of fact, unemotional demeanor, which is not to say that he did not appear interested or compassionate. He seemed weary from hard work and I took that to explain his somewhat tepid responses to what I thought were seriously disturbing incidents, dreams and thoughts that I had suffered since my hospital stay. I went on for ten or so minutes about the newest of my fears and concerns about what was happening to me. In the hospital most of the doctor's visits had been brief and he primarily talked to me about the biochemical aspects of schizophrenia and how that related to the vitamins and diet. I wanted him to understand what was at the heart of my concerns which was whether my perceptions were accurate and also about my bigger concern that if my perceptions couldn't be trusted then how could I know anything was real? If that were the case, how was I to know what would happen in the next instant or in fact what would happen to me in eternity? It was a frightening prospect and in some measure my reason for sharing my fears was due to a faint hope that the doctor would be able to point out some fallacy in my thinking that would undercut my logic and calm my fears.

When he stopped me I was certain he would have the answer I was seeking. Instead he began to question me about my diet. How much bread was I eating and was I sure it was one hundred percent whole wheat and not just the kind they call wheat bread which often had sugar in it? It was okay to eat a little lettuce as long as the salad dressing didn't have sugar in it. Better he said to use vinegar or just lemon and oil instead of a bottled dressing. He went on to increase the amount of vitamins I was taking. He went over with me again, as had in the hospital, about eating several small meals a day and being sure that I was getting one hundred protein grams each day.

I was a bit disoriented when he ended the interview in less than half an hour. I had assumed that the megavitamin therapy and the Salzaar high protein diet were to help stabilize my overwrought system but I had expected the appointments would be used to work through and solve some of the issues that caused me the fear. I was even more confused when the receptionist informed me that my next appointment would be a month away. I was still so mentally foggy and physically numbed from the Mellaril that the pang of panic that tried to grab hold of me when I heard that particular piece of information barely registered. I think I walked out of the office staring down at the appointment card with my mouth open, which probably helped me to fit right in with the rest of the patients that were coming and going from the psychiatrist's office. I took the card from my pocket again on the elevator and kept checking it all the way down. I was shocked that the doctor had sent me back out into the world to gut it out for a full month. I was bewildered that he hadn't wanted to talk with me at all about the things that caused my panic attacks. Thinking over those things did distract me from worrying too much on the way down in the elevator that when the doors opened I would have descended into hell or some other alternate reality.

I made it through the first month. I stayed very close to my parents' home in Lawrenceville where I was living. I don't think I did much besides sit and stare. My obsessions with reality and evil were with me but the anti psychotic drug still had me so knocked out that the fear was usually manageable. My side trips into the panic zone were less excruciating and they were sometimes shorter in duration. The attacks themselves would now tire me so much that I would sometimes fall asleep rather than spin out of control for hours or days like I had done previously.

On my second trip to Doctor Ward's office I noticed a familiar face. I looked up from my chair in the waiting room just in time to catch a side glimpse of his profile as he slipped out into the hallway. I believed he had noticed me also. He was a high school classmate that had run in a different circle than my own. He was Sicilian kid who had hung out with what we referred to as "necks." A lot of them wore big high fancy monogrammed shirt collars along with a kind of all black uniform of expensive pants, sweaters and shiny black shoes known as "featherweights." The word "neck" I speculated had come from the way they strutted peacock like around the school with their heads stretched unnaturally high as if to appear more "alpha' or at least to keep their chins up off of their three-inch-high shirt collars.

Although this particular kid dressed the part, he was actually a mild mannered somewhat shy person, unlike many of the mob wannabe troublemakers that shared the same dress code. I mentioned to Doctor Ward that I had seen a classmate. He told me quite briefly and without any specifics that he was seeing another Vietnam veteran who was also having some problems that had started during the war. This was the very first time that I had even considered the idea that my war experience could have anything to do with what I was going through. "How could it?" I thought dismissively. I didn't have it so bad compared to lots of people and as far as I was concerned I had left the war and anything else to do with the military behind me as quickly as I could.

I rarely spoke of my service unless someone specifically asked me about it. A couple of people at the college had questioned me one day in the Student

Center. I had answered a few of their questions under the belief that they had genuinely wanted to learn about what had really gone on there which, from my perspective, was far, far removed from the way it was being reported in the news. Suddenly their body language changed abruptly and they asked accusingly, "Did 'ja kill anybody?" I told them, "No, not today…yet," and I got up and walked away.

Just before the worst of my breakdown I had gone to a gathering in Cadwalder Park with some old high school friends. One of the speakers that day was Dan McKeen. Dan had been a junior when I was a senior at Ewing High. He had been president of the student body at Kent State University in Ohio when the National Guardsmen had tragically opened fire on the students, some of whom had been attacking them with rocks and bottles. I listened intently as he used Kent State as a launching point to attack the war, soldiers like me, the government and all manner of social imbalance and injustice. I was impressed by his passion and commitment but I was also struck by how passionately he put out complete misinformation about the war. There were things about which I had firsthand knowledge which were completely contrary to what Dan apparently believed.

Between my first and second visits with Dr. Ward my high school friend David came by to see me. I mentioned that I had seen the Sicilian kid from Ewing High School. David told me that the kid had been an infantryman in Vietnam. After he came home his father had thought it would be funny to sneak up behind him and make a loud gunshot like noise by slamming a book against the wall or something, which he did while his son was shaving and the son had whirled instinctively and slashed the father's face with a straight razor. I later learned through Doctor Ward that the kid was getting some help through the Veteran's Administration, including his medications. Somehow and I honestly don't remember at all how, as a result of this information some dots were connected and I ended up going for an interview with a state veterans service officer who helped me to initiate a claim with the U.S. Veterans Administration.

The service officer turned out to be the dad of a good friend from seventh grade, Rex Lau. Rex had been one of the two kids in seventh grade whose parents allowed them to wear a black leather motorcycle jacket and engineer boots to school. Rex's thick black hair was always perfectly coifed into a double DA, or duck's ass hairdo. Most nights in seventh grade Rex and I walked three or four miles home from Fisher Junior High School together as we almost always missed the bus because we both had detention, Rex for trying to buck the system like the delinquent James Dean character he was trying to play and me for not being able to refrain from getting a laugh in class for some smartass humorous commentary on the mistakes or other failings of my teachers. His dad told me Rex had become an artist and was living in Kansas City.

CHAPTER 12
Newark

As a result of my meeting with Mr. Lau, he filed a claim for me with the Veterans Administration and I began to make a series of train trips to Newark, New Jersey for an endless number of medical and psychiatric evaluations. I would park my Volkswagen at the Trenton train station where I would take either the New Jersey Transit line or the Amtrak to Newark. The ride was about an hour and along the way my mind would sometimes run away with fearful ideas. A frequent one was that the train would never stop. I would become afraid that it would continue to ride on for hours, days or even forever. I would anxiously fantasize that the other passengers on the train were all plants, in on the conspiracy. There was no point in asking anyone if it seemed like the train had been traveling much longer than it should be as I knew they would all just feign ignorance.

To save the dollar fifty in cab fare I would usually walk twenty minutes through Newark from the train station to the Veterans Administration building. Looking back, I remember encountering situations and people along the way which probably could have triggered some fear or anxiety in me, but there was none. I discovered through psychotherapy quite some time later that my fears were usually a reaction to my own thoughts. Interestingly, in real situations I was as calm as could be and often calmer and much better equipped to deal with situations that most other people.

One day on the train a man got into a loud and aggressive confrontation with a conductor. The man was likely inebriated and was certainly highly agitated. The conductor was escalating the situation by raising his voice to match the level of the man and had adopted an unyielding if not aggressive physical stance. The passenger was highly disturbed and began making physical threats to the point where I could see the conductor realized he might be in over his head. The conductor began to try to move very slightly backwards in the aisle and as he did the wild eyed passenger moved forward waving his arms and threatening to thrash the conductor. As the conductor flinched backward I came out of my seat and darted between them.

I got directly in front of the passenger, slowly reached out and placed my hand on his shoulder and looked him directly in the face. "You don't want to do this," I said, as quietly and convincingly as I could.

"Oh Yeah? Why's Zat?" he said, wildly.

"Because," I said calmly as I tried to look him in the eye, "you don't really want to hurt anybody."

Finally, he looked into my eyes and after a long moment he dropped his head and muttered. "No, I don't." With that he dropped his shoulders and the tension went out of his body. He simply returned to his seat and was quietly lost in thought for the rest of the time he was on the train.

Over the next many years I would intervene or help in a number of situations including muggings and fires that caused other people to freeze or flee. It would be a very long time before I would understand how deeply what I had experienced in the war had affected me. I did not suspect at the time that the fact that I reacted differently on the train that day had anything to do with having been in combat. Neither did I yet begin to suspect that the disturbances in my mind or more importantly my inability to live with my own feelings had anything at all to do with war.

I went through several batteries of evaluations at the V.A. including a particularly humiliating and frustrating interview with an aging psychiatrist who had a strong Scandinavian accent. "I'm just a pencil pusher." He told me. "You just tell me what you want to and I just write it down," he said. He tossed his pencil onto his desk indicating that he wasn't going to put forth any effort in trying to help me find a way to describe the anxiety he had asked me about.

While waiting at length in the Psychiatric and Neurological Clinic reception area prior to my interview, I had already begun to get the impression that I had not arrived at "help central." The lobby or waiting area was just a section of a much longer hallway and was defined only by a number of late 1940's era chairs and couches that lined both walls for a short distance. The aging furniture had dingy chrome frames and vinyl covered cushions. The vinyl coverings had a vague marbled pattern and half of them were dull grayish green and the other half non-matching faded maroon, possibly mirroring the lack of any kind of coordination in the rest of department. Unlike most other departments which had dedicated waiting rooms, the P&N section had its clients sit in a highly trafficked hallway showing a lack of sensitivity to the anxiety levels or self consciousness among its patients to say nothing of consideration for privacy or confidentiality issues.

Next to me was a man who kept turning his head sharply toward the side and almost shouting, "What time is it?" The first two times I indicated a large clock on the wall near the receptionist's station, which itself blocked nearly half of the hallway. "Twenty after eleven," I responded the first time and it was still twenty after eleven the second time he asked and I told him so. The third time, I merely pointed again toward the clock on the wall and after that I did my best to ignore him. There was a small black man sitting directly across from me rocking back and forth rhythmically as his finger traced line after line from the bible he was reading. His lips moved as he read but he was silent except that he would occasionally speak the word "Jesus" out loud and with great reverence.

On the couch next to the man with the bible was a man in a buttoned up winter jacket and gloves who sat stiffly and nearly perfectly still the entire time. In the chair next to the couch was a man in a wrinkled brown suit that kept bending forward to scratch his legs. There was another man on my side of the hallway who continually made reflexive movements with his legs and arms and his tongue kept darting from his mouth.

After quite some time a short thin gray haired and somewhat ancient woman, who looked not unlike a bag lady near the train station, came slowly down the center of the hallway. Her eyes were almost black and she stared unflinchingly straight ahead. Her face was pale and gray and she wore a graying full length lab coat with a very faded Veteran's Administration logo on the chest. I blinked when I caught sight of the worn blue name beneath the logo: Dr. Fluger! She passed by with out a word and without a single glance, even for a moment, side to side or up or down. The coat reached almost all the way to her ankles and she shuffled more than walked. The most prominent sound in the hallway for those moments was the ssssss ssssss of her soft soled shoes sliding along the worn out linoleum floor tiles.

As she passed by, six or seven of the nine or ten of us waiting in the hallway climbed silently out of their chairs and began to follow her. They followed along without a word, heads bowed as if in submission and they walked lethargically a good distance down the hallway and into a room on the right.

I had to ask so I slipped out of my chair and up to the receptionist's station where I inquired as to how much longer it would be before I was seen. The receptionist shrugged without looking up and I followed with the question I had really wanted to ask. With my thumb I indicated the direction that the old woman and the patients from the waiting room had gone. "What was that about?" Again without looking up at me the receptionist said matter-of-factly, "That's Dr. Fluger's eleven o'clock therapy group."

I looked up at the wall clock to see it now read twenty minutes to twelve. Between that and having gotten a glimpse of Dr. Fluger as she led her zombies to "group therapy," I was less surprised when it took another half an hour beyond the forty-five minutes I'd already waited to be seen. It was also a bit less of a surprise when the doctor I finally saw was old enough to be working towards a third retirement check or that he was as listless and uninterested in helping me as he stated quite frankly that he was.

After quite a number of months, quite possibly more than a year, the Veterans Administration adjudicated that I had a ten percent service connected disability for my "nervous condition." I would now be able to obtain my medications directly from the V.A. by mail...and would also receive twenty-eight dollars a month in disability compensation. Dr. Ward wrote back to the VA suggesting that ten percent seemed quite low and perhaps they hadn't understood the severity of my disability. The Veterans Administration did not respond to his appeal.

CHAPTER 13
Back to School

As I battled on through the schizophrenia symptoms, I continued to try to function. I attempted to go back to school at Trenton State College in September of 1970. My problems had been building through the previous spring semester but I'd held them in check to some extent and escaped with a B average before the real breakdown came in June. Now Dr. Ward's promise that the diet, the vitamin therapy and the drugs would stabilize me helped me gain enough confidence to try to continue my studies. I enrolled in Sociology of the Family, Existentialism, Intro to Psychology I and Principles of Economics II. In the fall of 1969 I had taken 15 credits and my grade point average was 3.75. Now in the fall of 1970 I took on a reduced load of only 12 credits. In spite of that my semester grade point average fell to 1.50. I got a B in Psychology and a C in Existentialism. I failed Sociology of the Family and earned a D in Principles of Economics II. At the time I chalked up the Economics grade to my condition as I had done well in Economics I in the previous spring semester and I usually excelled in mathematics related subjects.

In retrospect I see that the Macro Economics theories that were the focus of study in that second term were grounded in real logic or evidence about as much as the catechism answers I had been forced to learn in Catholic grammar school. Looking back at age sixty-seven at all the wildly opposing economic theories I have heard over the years from government officials and Wall Street strategists and watching their universal failure to be able to predict or control economic trends even sometimes a day or two ahead of time, I now realize no one really knows anything about the subject including the authors of the books we used. Like religious subjects, only blind faith in theories that failed to hold up under logical analysis could have quieted my confusion in that class. The sad part is that my anxiety about my mind not working properly and my fears that other people would recognize my Schizophrenia and shun me, caused me to remain silent and keep my many questions to myself.

I started the semester in a Physical Education class but I had to drop it almost immediately. Bouncing on a trampoline during one of the very first classes I was seized by chest pain. I fell to the canvas, dropped over the edge of the trampoline and almost crawled to the hallway outside the gym. My heart pounded wildly from side to side and the pain in my chest spread down both arms and down my legs. My vision began to white out. I leaned back against a wall, dropped to the floor in a sitting position and let my head droop on my chest. My pulse rate sped completely out of control. It was twenty minutes before I could get up without everything going black and my heart was still speeding as I made my way to the locker room, clutched my street clothes to me and somehow got to my car still wearing the gym clothes that were now soaked with sweat. I fell into the seat, set the backrest all the way down and remained there for some time staring up at the dome light. I had many, many such attacks over many, many years. Why I didn't call for help, why I didn't go right then and there to a hospital or at least to the college nurse's office I cannot explain.

Years later in Los Angeles I saw a doctor named Jim Blechman who attempted to help me with this condition. He asked me if when the attacks came I felt as if I might die. I told him that it did occur to me sometimes when they first hit but that I was probably more afraid that they would get me into the hospital and kill me there. Jim was an MD who had also been trained in Naturopathic and Homeopathic disciplines. He answered wryly that he thought that being killed by the hospital doctors was probably a realistic fear.

I had two more episodes before I saw Doctor Ward again about a week and a half later. By then I believed I had already begun to piece together a partial answer. I found the episodes usually came at a point when my mind was speeding ahead about something and that my blood sugar was also low from not eating enough or from over exertion. The Doctor seemed relatively unconcerned about the heart episodes but, I noted, he never showed a lot of concern about much of anything. His affect was generally so flat that I had begun to wonder about his ability to relate to the deep and terrifying feelings that I suffered from. In his usual matter of fact manner Dr. Ward prescribed a drug called Inderal that which he said was a "beta blocker," which meant nothing to me. He said it might slow down the heart rate under stress. He ordered an electrocardiogram, the results of which were essentially normal. The doctor also wrote a note which allowed me to drop the Phys Ed class.

The attacks went on for many years often lasting for six to eight hours and almost always at least three or four. They were always violent and always took me by surprise. The doctor took me off the beta blocker after a few months when it became evident that it had no effect on the problem. Dr. Ward said it was better to stop it as there were also risks to continuing on it.

I suspected that blood sugar levels to my brain were dropping and that my heart was racing to try to make up the difference. This made sense to me as my peripheral vision would usually decrease and I would start to "white out" just before the big jolt would slam my chest and my heart would run away and go off kilter. Sometimes it felt as if the right side of my heart was pounding and sometimes the opposite. I tried quickly drinking a glass of orange juice to raise my blood sugar level and then following immediately with some hamburger or other high protein food to keep the levels up. That tactic did seem to lessen the length of the attacks somewhat … sometimes.

I had a number of attacks while driving. On some of those occasions the heart palpitations hit first and I had to just gut out the pain until I could get off the freeway or pull to the side of a road. When the vision white out came first it was more difficult, but always I seemed to be able to maintain just enough eyesight and just enough control of the pain to manage the car until I could get to the side of the road and safety. I do believe that had I not been in combat before these episodes happened that I would surely have panicked and likely lost control of the car. There was a small but adamantine part of me that had kept me moving forward in otherwise overwhelming situations during my Vietnam tour. When the heart episodes would strike, my mind would initially go to fear and pain. My chest would hurt my arms and legs would ache. I would worry about crashing into the median or another car as my vision tunneled. I would feel as if I were about to lose consciousness but that tiny part of me would become ever so slightly clearer as it had in firefights and other desperate situations in the war, and I would maintain just enough control to get to safety.

Some several years later I learned that the anti-diabetic medicine I had been taking to help control my blood sugar levels was quickly and quietly removed from the market. The drug, prescribed by Dr. Ward, was known as DBI and was made by the Lily drug company. It was found to cause fatal toxic lactic acidosis. There was also some speculation about a suspected association with heart problems. I stopped taking the drug immediately on learning its life threatening side effects. Recently, while doing during research for this book, I learned that the drug Mellaril, which I was taking in very high doses when the heart condition began, is now associated with torsade de pointes-type arrhythmias, a potentially fatal polymorphic ventricular tachycardia, and sudden death. I continued to have attacks for many years and in fact am still vulnerable to them on occasion now, more than forty years later, although they take a milder and more controllable form.

Dr. Ward continued to see me once a month. The half hour sessions would begin with me attempting to convey to him the level of my fear and the situations in which it had been triggered. He would in short order interrupt and begin to question me in depth about my diet asking detailed questions about exactly what I was eating, how much and when.

The doctor, along with several leading researchers of that time, was convinced that schizophrenia was the result of a chemical imbalance in the brain and that blood sugar levels were critical in controlling the disease symptoms. On a couple of occasions Dr. Ward would also adjust the levels of the antipsychotic drug or add to the massive amounts of the "Mega- vitamins" that I took to try keep the symptoms at bay.

Over the next six months the severity and frequency of my torment did begin to lessen. Although I was by no means free of psychotic symptoms, I did have less moment to moment dread and was therefore more than willing to comply with the diet or for that matter just about anything that was asked of me. At times I still feared that I was already dead and was living in some kind of hell that was orchestrated by my own unconscious and out of control thoughts or by some invisible great and evil mind that was the source of all things. The anxiety levels were lower however, and the really intense terror attacks had all but disappeared, so I faithfully ate one hundred grams of protein a day and as little refined carbohydrate as possible.

An egg has six grams of protein and a quarter pound of hamburger has about ten grams so ingesting one hundred grams each day is a nearly full time job and not inexpensive either. I weighed about one hundred twenty-five or so pounds at that time having returned from the Vietnam War weighing one hundred twelve as a result of my multiple bouts with gastroenteritis and malaria symptoms. In spite of the enormous amount of food I was eating my weight stayed low. My body in general was poorly toned and in fact my sense of my body was very poor. My skin sensitivity and my sense of touch were dulled to a great extent as a result of the antipsychotic medications which are designed to block nerve impulses. I did not get much exercise in part because I became afraid to cause one of the painful and frightening heart spells and in part because by limiting my activities I encountered less people, places and things which could trigger a terror attack and the days of obsessive mental spinning that followed as I would try convince myself I wasn't really in hell.

While my days became somewhat more manageable, I began to have a very disturbing nightmare that mirrored in a bizarre way an experience that I had had in Vietnam. In the dream I would be in a bunker at night with one other MP. We came under attack by a much larger force. The enemy soldiers were dressed in glowing white outfits and the entire scene was bathed in a purple haze as if lit from within by ultraviolet black lights. My rifle would jam, as it actually had in a similar combat crisis during the war, and I would frantically clear the jam and begin firing with extraordinary accuracy, as if guided by some supernatural force. I could see my rounds penetrating the bodies of the enemy soldiers, but without any effect! I could see where the rounds entered, but the glowing white material of their uniforms remained unbroken. The enemy soldiers continued to advance, running through barbed wire as if it wasn't there, then through the sandbag walls of the bunker I was in and finally they ran right through my own body as if I wasn't there at all. At that point I would always awake from the dream gasping for breath, my body arching upward off the mattress in massive spasm. Once awake I could not remember the specifics of the dream. At the time I remembered only that it had something to do with a firefight and that I awoke in bathed in terror.

I told Dr. Ward about the repeating nightmare and my inability to remember it. He nodded calmly and explained in his usual unemotional manner that he believed my problem had started in the war zone and that I had begun drinking heavily to suppress the symptoms. He did not indicate that he felt there was anything to be gained by exploring those dreams and memories and he rather shifted the subject to whether I was buttering my bread or whether the peanut butter I was using had sugar in it.

Looking back, I can see that the dream caused me to feel as if my body were dissolving and contained levels of intense fear that clearly matched the feelings I had during the panic or terror attacks I'd previously had during waking hours. I did not have the mental capacity to make that connection at the time. In addition, Dr. Ward had thus far provided the only answers which had relieved my horror and which had given me any hope of being saved from the black evil that had hounded me toward the eternal and terrifying void that lay at the edge of all my thoughts. To question the good doctor would have been to give up the only hope I had so I merely went about eating small high protein meals several times a day and trying to make it to school and back for at least most of my classes.

I asked the doctor how long it would take to get better. He did not answer me directly. Instead he told me in his usual less than emotional manner about what happens, "if you take an axe and gash a tree." He went on to explain that for the few days after you "gash" the tree, the wound is fresh and sap runs from it. He said if you leave it alone and don't re-injure the tree the sap will harden and within a few weeks a new surface will form and although the wound is still evident, it is beginning to heal. He said in a year's time new bark will have begun to grow over the injury and in five years time you would probably not even notice that the tree had been hurt. I took his little parable to mean two things. That some degree of healing would occur within the natural order of things and that my recovery would be a long and gradual process. Without saying it in so many words he was telling me not to expect to be suddenly "cured" or relieved completely of my symptoms at any point in time.

CHAPTER 14
Relapse

Suddenly I felt myself tumbling uncontrollably back down the rocky slope I had worked so hard at climbing for the last several months. My fear levels became unmanageable. I was readmitted to Mercer Hospital as an emergency.

The doctor claimed that it was a positive sign. He said that the heavy medication I had been on along with the diet and megavitamin therapy had allowed my system to heal or rejuvenate sufficiently for me to now have the strength to begin to deal with some of the hidden feelings and problems that had caused the breakdown. He maintained that I was actually stronger, strong enough to handle feelings that had been hidden from me because they were previously too overwhelming to deal with consciously.

That was emphatically not what it felt like to me. What it felt like was that I had tried desperately for six months to delude myself into believing that the world was a real and workable place and that my terror came from something ultimately manageable. Now the rug had been pulled out for good and I could no longer doubt that I was dead, that everything I saw, heard or felt was an illusion and that it was only a matter of time until the black evil that was the void revealed itself fully to me.

I remember the doctor calmly sitting by my bed in the hospital and telling me not to worry. He was going to change me to a different medicine and that soon every thing would feel better again. Just a few moments later I overheard him talking to my parents in the hospital hallway. He told them quite oppositely that they should prepare themselves for the possibility that I might have to be institutionalized permanently at some point.

Hearing conflicting messages like that would probably make a normal person quite anxious. The idea of ending up in the state mental hospital should probably have filled me with fear but I wasn't particularly afraid of that. What I was afraid of was being sent from one hell to the next for all time with each hell being worse than the one before.

I began again to obsessively search my thoughts for some rational explanation of what I was going through, some reason to believe that the world and my body were real and that I was merely sick. Finding none I began to look for an escape. I had eliminated suicide some time before for two reasons.

I believed that if the world was ultimately illusory I would simply wake up in an alternate reality or illusion, like shifting from one dream into another. Secondly, if the world was a game being played on me by the great evil in the void, then I was afraid I would be punished for trying to escape by killing myself and would surely end up worse off than I was at the moment.

I spent my waking hours in the hospital thinking deeply and obsessively about my true nature and about the nature of the outside world. One of the tenets of science as I understood it in those days was that in order for matter to be real, the smallest component parts of it had to be indivisible. In the days when the physical sciences were looked up to for the ultimate answers, many believed that molecules, then ultimately atoms were the smallest particles of matter. Since the advent of Quantum Physics and since scientists have learned to split the atom, that view that matter exists on its own as the ultimate "stuff" of the universe has been discredited. Now scientists have found multiple subatomic particles, some called Quarks, and within them other semi-material energy/particles which seem to flip in and out of existence. The Quantum physicists have gone so far as to say that nothing exists objectively as everything in what we call physical reality is in some way affected by the mind of the observer.

As I lay in the hospital, my body was bathed in fear in spite of the months of medications, restricted diet and massive doses of the megavitamins. I now had little hope that I would be saved by the doctors or by the "Goodness" of the universe. I was living through terror that was almost unimaginable. I had begun to accept my worst fear that at it's heart the world was evil and even that possibility didn't horrify me as much as it had. What truly sent me toward despair was the idea that the horror, the awful terrifying feelings in me would go on eternally, without end.

CHAPTER 15
Nirvana

I had been lying awake for many hours. I began by thinking about how the lens in our eye inverts everything. The images are actually upside down on our retinas. Our brain apparently then takes that information and projects a righted image somewhere else in our brain or our mind or wherever it is that the "me" I was looking for resides. Our experience seems so direct and immediate yet according to the physical sciences we learned in high school all of our experience takes place somewhere inside the brain and only after a time lapse, however short, from the real event. If we only know reality indirectly, that is after the brain has reorganized and interpreted information, how then do we know the difference between what is real in the outside world and images that we imagine, like those in dreams?

For days I had been trying to find one single thing that I could point to and say for certain that it was "real," that it had a physical objectifiable existence. Try as I did I found it impossible to prove for certain that anything existed outside my mind. I then set out to investigate the opposite, that non existence was possible. Previously when I had tried to visualize nonexistence, an empty void, a nothing; I'd had the vision of the Big Bang as it was later described by Hawking.

Einstein had described reality in terms of "relativity," claiming that matter, energy, space and time were not separate entities, but existed only in relationship to each other. I began to think about time in the same way that I had thought about matter. "What was time?" I thought. Did time contain smaller component parts that were indivisible? I couldn't see how. Where did one second end and another begin? In fact, how could we even draw a line between a nanosecond and an hour or a year? How could we separate any single second and eternity for that matter? Maybe time was merely an organizing concept in our minds. Maybe everything existed all at one time in one place and our minds just perceived things spread out over time and space.

All of these machinations of course were motivated by a desperate need to believe there was some escape from the hell I was living in, some final end to it all. I didn't believe suicide would kill my mind. My Big Bang visions had defeated hopes that the universe might at some point decay into nothing or annihilate itself because it seemed the "nothing" would contain the same potential energy and cause the universe to "bang" back into existence again and again.

What about the time element? What if one could find the blurry area between one second and another, between a second and eternity? What if I could find a state of mind with no time? If mind was who I really was, maybe I could escape into some timeless moment within it, a place of peace or even some kind of non experience. I could stay there and let the rest of the evil world go on without me. That, I thought, had to be the nirvana the religious people were talking about.

I remembered that the Maharishi Mahesh Yogi had come to our campus at Trenton State College. The yogi had advertised in the student newspaper ahead of time telling people that they had to bring offerings to receive a mantra that was especially created for them. The required offerings were specific. The devotees were to bring a flower, a clean white handkerchief, each which was symbolic of beauty and purity of their devotion to the guru, and a third item, twenty bucks, which as far as I was concerned was the whole point of the exercise. Twenty dollars was serious cash to a State College student and a couple hundred students contributing twenty dollars would make a good take for a couple hours of guru work.

For your twenty bucks and your symbolic tributes acknowledging the alleged beauty and purity of your connection to the guru, you were given a secret mantra, to repeat over and over in your meditations. The secret word or phrase supposedly would attract forces of whatever form of happiness you were seeking. You weren't allowed to share your secret mantra with anyone else. This made everyone suspicious that the mantras being sold might not be as individualized as advertised. Several of the students I knew took the chance of revealing their mantras to other devotees risking losing their shots at acquiring Nirvana or a Rolls Royce automobile like the one the guru was driven around in.

I didn't know a guru from a lama or a lama from a llama or whether it was the former or the latter that had a hump, or was it two? I knew nothing about formal meditation technique but when I remembered the guru I began to wonder if nirvana was simply a state of mind. Maybe the answer was to escape the outside world by retreating fully into the mind. My mental efforts slowly turned to finding a quiet place between two seconds or moments in time where I could stay ...forever.

As I tried to envision the nothingness and timelessness that I sought, I became more and more aware of my anxieties, anguish and of the knotted pain in my body. I'd had so many terror attacks and now more recently the heart episodes that my body had begun to coil inward on itself in a kind or anticipatory or defensive posture. For the moment my mind was only awake enough to recognize the pain, not the associations or anxieties that caused the spasms and constrictions. I had normalized pain for so long that I had limited awareness of it even when I tried to meditate. Pain had become an integral but unconscious part of my sense of self.

The more I struggled to quiet my mind in order to escape my body and the world, the more it made me aware of my body and my surroundings. I would revisit this concept much later in life and recognize it as real insight. In my then current state of agitation and hopeless fear it was merely another disappointment in my quest to believe there was an escape from the awful fear filled world I lived in.

Failing to find a place within of emptiness or nothingness, I shifted my efforts to imagining a place so perfect that my mind would want to freeze itself there forever. There was an image that had given me a real sense of peace when I first saw it. It was a Maxfield Parish painting of an idyllic wooded spot up above a creek or river bed that ran through the valley below. There were a few slender youthful looking men and women lounging and looking out from a kind of marble or granite veranda that had Grecian columns. Although I could not fully visualize the painting, just recalling the first time I saw it calmed me and made my body begin to relax. I could feel the warmth of the sun on the lightly clothed subjects in the painting almost as if their bodies were my own. I could hear the gentle flow of the stream below and the breeze softly blowing through forest above. There was something about the predominately blue tones in the painting and the sense of bright white sunlight that spoke to something deep within me as well.

Although I could imagine a brief sense of relief or calmness when I thought of the Parrish painting, there was still a powerful feeling of terror at the edge of my thoughts. Relaxing into the idea of nirvana would mean letting go of the sense of control that I so desperately sought to maintain. I was still very much afraid that the world could not be trusted. I had seen and experienced many things as a child and later during the war that made me seriously doubt the idea of a good and loving God. I also lacked trust in my own senses. I was afraid that everything I experienced including my own body might have no more reality than an ordinary dream, or more to the point in my case… a nightmare.

A part of me was constantly monitoring and dissecting every experience including my own thoughts. That hypervigilant part of me was determined to analyze and interpret everything I encountered looking for hidden messages or meanings that would confirm or negate my fears about the world and about my fate in eternity. Yet another level of my suspicions were devoted to keeping watch over my own thoughts, passing judgment on whether those thoughts were my own true and accurate perceptions or conversely whether they had infiltrated unnoticed across the unseen border between my own mind and the great mind of evil that was in the void beyond. It was exhausting work.

While I was in the hospital I overheard my father complaining to Doctor Ward that I would sometimes just sit and stare. My body was exhausted from my mental machinations, from fear attacks, heart episodes and from the side effects of the drugs. My mind was truly as active as any one person's could be, hyperactive really, as I spent all of my waking hours trying to sort out the nature and destiny of the world as well as my own fate within it. From the outside however, I apparently appeared virtually catatonic at times.

From time to time over the next many months of my disease I would often sit and stare. Sometimes I was merely experiencing fear or anxiety triggered by something someone said or by something on the TV or a song on the radio which I suspected had a hidden message directed at me. At other times I was trying fervently to find that timeless place in my mind where I could escape forever into the heavenly realm.

Sometime during that period I watched a TV pilot for Rod Serling's Night Gallery. One of the three stories in the episode was about a man who felt so absorbed by a painting in an art gallery that he began to believe he could project himself into it and ultimately find eternal peace within the pastoral scene it depicted. In the climax of the story the man becomes pursued by his tormenters in the outside world, breaks into the museum at night and focuses all his mental powers on an attempt to escape once and for all into the painting.

The next day gallery visitors pass by a painting that sure enough has an image of the man within it that wasn't there previously. The rub is that he has escaped not into the impressionistic pastoral scene that he sought but rather has committed himself for all time into a Hieronymus Bosch like scene of a hellish nightmare in which he is being tormented by sadistic demons. That pretty accurately encapsulated the fears I lived with on a daily basis. Without much further internal discussion I immediately abandoned all further efforts at escaping permanently into the uncharted realm of timeless mind.

It was actually a brief sense of timeless space or eternity that had precipitated my recent relapse and landed me back in Mercer Hospital. I had been driving in my Volkswagen. I came to a stretch of road through a wooded area where there was nothing man-made except the roadway itself. There was a James Taylor song playing on the radio. As I entered a curve where I left behind all the man made and commercial buildings, telephone poles and the like, I was struck by the purity of the completely natural surrounding I was now in. On the radio, Taylor sang the word's "'cause I could feel it…walkin' down a country road." At that very moment I felt as if my individual body dissolved completely and that I was without any separation between myself and the rest of the natural world. There was a sense of light coming from within everything and a feeling that everything down to each leaf of each tree had a sentience or intelligence within it. I momentarily felt a sense of fleeting bliss but that quickly gave way to the overwhelming fear and terror of being completely naked and vulnerable to a universe or intelligence that I had no reason to trust and every reason to believe was ultimately evil, based on my past experiences.

I was now more convinced than ever that the world was composed of a much softer or more fluid reality than most people imagined. I was still terrified that the ordering principle beneath it was something far removed from the loving kindness of God that gave most people the comfort that I was so devastatingly missing.

Dr. Ward continued to assure me that my problem was biochemical and that some adjustments to the diet and medications would help. I was however inconsolably terrified. Eternity facing supernatural evil was the problem I faced.

CHAPTER 16
Dr. Jane Rittmayer

I was in the hospital for two full weeks. They fed me the Salzaar high protein diet and continued me on the Megavitamins. Dr. Ward added more niacinamide to the regimen. I was now taking nine grams of B-3 a day, or eighteen 500 milligram choker tablets a day. I took twelve 500 milligram vitamin C tablets, 3 1000 I.U. capsules of Vitamin E, as well as tablets of Vitamin B-6 or pyrodoxine. I was also still on the DBI capsules to control my high insulin levels which in the doctor's view contributed to the low blood sugar levels in the brain. Mellaril, or thioridizine, my antipsychotic medication, was replaced by Quide, generically piperacetazine. I was released from the hospital feeling not a lot better than when I had gone in.

Over the next few months, I did find some relief from some of the side effects of the Mellaril. Quide was touted as more powerful and as a result I was able to take much smaller doses than with the previous prescriptions. On the new drug I eventually began to lose some of the dull feeling I had suffered for so many months. Some ability to taste returned and I began to feel, for example, air moving over my arms when there was a breeze.

The aim of the drugs in the antipsychotic class is to relieve symptoms including the overwhelming anxieties and fear that schizophrenics can experience. This is done by blocking neural impulses. The impulses are blocked non-selectively so while emotional signals are inhibited from reaching centers in the brain, sensory impulses coming from stimuli in the outside world are also blocked. This has the decidedly negative effect of further separating the sufferer from the very world that he or she fears is unreal.

From a doctor's point of view, the symptomatic behaviors may be diminished. The sufferer however may experience vastly increased symptoms subjectively as a result of the reduced capacity to relate to his or her surroundings. The ability to communicate what is going on inside is also lessened by the dulling effects and by the fact that the drugs cause extreme drowsiness and lethargy in many cases. In addition, the antipsychotic drugs or major tranquilizers as they are known cause a number of frightening side effect ranging from blindness and ejaculation inhibition to a wide range of involuntary tics and spasms. These spasms often cause other people to give sufferers a wide berth, increasing their sense of isolation.

Switching from Mellaril to Quide did seem to lessen the feeling of gray dullness and exhaustion that had plagued me. As my senses slowly returned, my bodily feelings and even my vision seemed sharper. Before long and without warning however, the new drug would also cause a very serious physical side effect that would land me back in the hospital once again, this time in critical condition.

I saw Dr. Ward at his office a short distance from the hospital within a couple of weeks of my discharge. As soon as I got there he introduced me to another doctor who he said would be working with me. Without any further explanation, Dr. Jane Rittmayer then led me from Dr. Ward's office to another room around a corner and down the hall within the same office suite. I remember being pleased that the new office had a window facing me from behind the doctor's desk. Dr. Ward's office had always seemed gloomy.

I later came to understand that Dr. Rittmayer was a psychologist not a medical doctor and that her doctorate from Rutgers University was an EdD not an M.D. or PhD. No one shared this with me at the time and I would have had little understanding about such things anyway. At that point I had little hope that anyone could help me and in fact had recurring and frequent doubts that anyone or anything I encountered was anything more than a dream or a mental projection of some kind.

The new doctor was pleasant as well as somewhat pleasant looking. She appeared to be in her mid thirties and she dressed in conservatively tailored skirted business suits worn over a blouse or sweater. She kept a notepad nearby, but unlike Dr. Ward, who seldom looked up from either his tea or the notes he was making about my diet and medication regimen, the new doctor's body language clearly indicated that she had an interest in what I might say. Initially, that wasn't much. I was still in a great deal of fear and I was still in doubt about everything I experienced. I wasn't yet sure whether or not this new doctor was yet another trick being played on me by the Devil to lead me into the next chamber of hell that he had waiting for me.

There were long periods during the first session with Dr. Rittmayer during which neither of us said much of anything. I was very much taken by surprise by the sudden shift in doctors and was disoriented. Confused and unsure is actually a better description than disoriented. There was precious little in my life situation at that point by which I could orient myself. Although I had been attempting to continue at college, I had no direction there, no job or career path, few friends and no real home. I was staying at my parents' house where they had moved from our old neighborhood while I was away in the Army. Not expecting me to return home they had understandingly bought a home that didn't provide a real adequate room for someone my age. I had precious little to orient myself by with the exception of my mental state at any particular moment. On any given day that ranged from moderately debilitating anxiety through fear and infinite paranoia to naked terror. Throughout most of each day close to one hundred percent of my limited energy was focused on what level of fear I was experiencing and on trying desperately to avoid any thoughts or situations which could set off an attack that would worsen those fears.

Dr. Rittmayer suggested that I see her each week instead of waiting the month long interval as I had with Dr. Ward. Early in our second session she asked me if I would be willing to take a test with her as a way to help her understand what kind of resources I might have to work with and as a way for her to get to know me. I agreed and she began to administer a battery of items all of which I found quite easy to dispatch quickly and without a great deal of effort. Dr. Riitmayer showed me a series of cards and asked me a number of questions while using a stopwatch to keep careful track of the time it took me to answer. She made frequent notes on a score sheet. The doctor seemed more and more excited as the testing progressed. Some questions were mathematics problems; some were picture cards on which some missing element had to be identified. The final set was a series of knowledge questions. The very last question on the test was who wrote "Faust?"

I had no idea on that one but Doctor Rittmayer excitedly told me that I had answered every other question on the test correctly and had received the maximum points allowable for answering quickly. When I asked her what kind of a test it was she told me it was a WAIS or Wechsler Adult Intelligence Scale or IQ Test. She told me I had tested very high on the scale. When I asked her how high, she hesitated momentarily and then threw out the number 150 with a little turn of her head as if to indicate "approximately." I didn't know at that time that 150 was the highest achievable score on that particular test. I knew I had previously tested well beyond my ability to perform in school; something which had caused me to be harangued on an almost weekly basis by the Ewing High school guidance counselor, Gladys Jensen.

When the session ended Dr. Rittmayer accompanied me as far as Dr. Ward's office where she ushered me inside and announced quite proudly that I had tested very high in the superior range on the IQ scale. Dr. Ward looked as surprised as I had seen him look as he glanced back and forth between Dr. Rittmayer and me. I had the sense that my new doctor was patting herself on the back a bit for discovering something that had been missed by Dr. Ward.

The next week's session started slowly again. Dr. Rittmayer mentioned the testing and said that it was somewhat encouraging. She said while very bright people sometimes struggle in recovery because they can get lost in deep or complex thinking, that on the other hand they had resources available to help sort things out that other schizophrenics did not have.

I recoiled at the word schizophrenic. The session ground to a halt. After a long period of silence, I began to tell the doctor, much as I had tried previously to tell Doctors Ward and Silverston about my symptoms and then quite frankly about my fear that I was not suffering from a disease at all but was trapped in some dreamlike reality which was controlled by some great evil.

Dr. Rittmayer seemed to be listening much more intently than anyone had before, so I continued on and confessed quite frankly my horrible fear that some of the terrible things that had happened around me may have been caused by my own thoughts. The fear built in me as I spoke but I pressed on needing so desperately to be taken seriously. The other doctors had been dismissive about most of my symptoms and especially about my terrifying view of the world. They had always acted as if these things were essentially irrelevant to their own system of dealing with me which was to try to alter my brain and body chemistry.

As I continued on, my fear became so intense that it was the only thing I could feel. I lost any sense of body and felt only terror. This was the naked exposure to the unseen world that I fought so hard from moment to moment to avoid. The sense was one of having only an emotional or mental body and no real physical existence. It was what I most feared. There were no boundaries... no center and no edges to my existence. I felt as if I could dissolve and that the outside world could dissolve around me and things could spin out of control much like they do in a nightmare only with no hope of ever being able to awake from the horror.

I froze and became silent. Now that I had fully confided my deepest fear I became afraid that the doctor might disappear from the room or worse transform into some version of the Devil.

She studied me for a moment and then began to speak tentatively. "I went through something similar, Tom," she said softly. "Mine was not exactly the same but similar."

I sat silently. She went on. "I thought I was a witch," she told me. "I believed my thoughts were causing people to die or have other bad things happen to them."

I was dumbstruck. This was the first time anyone had said anything that made me feel like they could remotely understand what I was going through. Not only could she understand but she could really understand! She knew what I was talking about!

I did not move or speak. My wheels were turning, assessing whether this was a trick or something quite different. I had noticed a hidden shiver in Dr. Rittmayer when she spoke of being a witch and causing harm with her thoughts. I knew by her tone she wasn't making it up. My experience with her confident manner over the last couple of weeks also told me she was no longer under a spell like I was for I knew all too well she would have been unable to hide that kind of terror for this long.

There was still a great deal of mistrust in me, but something changed at that very moment when Dr. Jane shared her own history with me. What changed was that I now had hope.

CHAPTER 17
Criminal Justice

Except during the periods of hospitalization, I continued to try to function, with varying results. I switched to part time attendance at Trenton State College, enrolling in night classes. I was required to declare a major and chose Criminal Justice. I had no idea what career track I might take if I could complete my degree and didn't have a particular interest in police work in spite of the fact that I had been a Military Policeman in the Army. At that point in my disease process my sense of self was nearly nonexistent and my thoughts were deeply and thoroughly tied up in my obsessions. Thoughts of the future were entirely about how I would deal with an eternity in fear and pain. I had no real ability to realistically project interest in any area so I more or less chose criminology by default. Because of my past training as an MP, I thought it might at least be something I could relate to academically.

I became somewhat more comfortable and able to communicate in my therapy sessions. I began to function minimally in some other limited settings. I was still extremely withdrawn and insecure in the school environment. I took a photography course in which the first major assignment was to take an outdoor portrait. I was too withdrawn and shy to ask any of my classmates or even a family member to pose so I took photos of my parents' mixed breed miniature collie, Andy. The photography professor ridiculed me and berated me in class for being stupid enough to waste film on a dog. I don't remember anything else from that class for the rest of the semester.

Most of my classes at that point were night classes in my new major, Criminal Justice. It was rare in those days for policemen to have college degrees and many of my classmates in the night classes were older men, mostly prison guards, looking to advance their careers. I felt as if I fit in even less with these people than I had with the younger kids in day classes.

We did an exercise in one of the classes to demonstrate the difficulties in making split second decisions about firing a weapon at suspects. The teacher handed out clackers to each student and showed slides of ambivalent situations we might confront in law enforcement. There were slides of people who looked innocent but in the next slide revealed a concealed weapon. There were other people who looked dangerous but in the second slide were shown to be harmless. We were supposed to click the clackers as soon as the first slide appeared if we would have fired in that situation. The second slide showed whether we had been correct. The point of the exercise was not to hone perceptions or skills but to show the difficulty of making correct spit second decisions under pressure.

The teacher was a young New York City patrolman who held a Master's degree in Criminal Justice. I felt the exercise was being used to justify wrong decisions or "bad shootings" as they were known in the trade. The mood became almost jocular among the middle aged prison guards and cops who began to joke and share stories about how they had been taught to carry an untraceable weapon to plant on a suspect in case they killed someone by accident. I became unbearably upset but held my silence for fear I would start into a panic attack if I tried to speak out. I was very likely the only person in the room who had actually killed other human beings, but at the time I didn't feel like a wizened veteran with experience to share but rather a mixed up, insecure adolescent who had never done anything right.

After that day I found it more and more difficult to concentrate in classes, my mind constantly running to obsessive thoughts about how I didn't think I could ever carry a gun again and about what in the hell I would do in the criminal justice field without being found out. Although the most extreme fear attacks were infrequent, the sense of unreality was with me almost constantly. My anxiety was high and my spirits were quite low. On the way across the athletic fields between the classrooms and the parking lot at nights, I began to stare up at the campus water tower and wonder what I would feel or think during the few seconds it would take to fall from the top to the ground and death.

CHAPTER 18
Working

After seeing Dr. Rittmayer for a time, I made an effort to try to support myself more fully. I was living in a tiny room in my parents' home in Lawrenceville, New Jersey. I had a small stipend of about one hundred sixty dollars a month from the G.I. Bill to help with college costs and a much smaller payment of twenty-eight dollars a month to compensate me for my service connected disability which was rated at only ten percent. I owned a 1970 Volkswagen Beetle which had replaced a 1969 model of the same car which I had rolled over driving home in an early morning alcoholic blackout just prior to my breakdown. I had purchased the first Volkswagen to replace a 1965 Chevrolet that I had also totaled during another alcoholic blackout. Both cars had been purchased from funds I saved during my tour of Vietnam but now that account was empty.

I applied for a job as a lineman at the downtown Trenton office of New Jersey Bell Telephone. I had no particular reason for choosing that company or that job except that I had heard it paid well. I was required to take a battery of aptitude and other tests after which I was given an appointment with a counselor from the personnel department. I sat quietly in front of her desk as she looked over the results of the testing. She looked back and forth between me and the papers on her desk several times before speaking. Finally, she gave me a questioning look.

"You did very well on the testing." She paused. "Why in the world would you want to be a lineman? "

I had no idea. "I just need a job," I told her.

She shook her head. "All right," she sighed as she put my papers in a folder and placed it to the side. "Someone will get in touch within about a week."

In far less time than that I got word that my application had been denied for medical reasons. On the application I had listed the reason for my recent hospitalizations as hypoglycemia or low blood sugar. This was the underlying cause of my mental and emotional problems according to Dr. Ward so I felt justified in reporting that rather than risk the stigma of mental problems. As it turned out hypoglycemia was enough on its own to cause me to be rejected. The telephone company wasn't willing to risk having me faint or become weak and fall from a telephone pole. I later learned from someone with a labor law background that the company was forbidden by law from rejecting me without having given me a physical examination themselves to verify my condition. I was told I could have had recourse but it wouldn't have mattered. My sense of self and of direction was so deteriorated that I could not have begun to fight with anyone about anything. The battle raging inside of me took essentially all I had to give. In the outside world I merely drifted along like a fallen tree branch in a muddy river.

Somehow I drifted into a job fair for veterans at the National Guard Armory in downtown Trenton. I was offered a job at the very first table I visited. Within seconds I was hired by the New Jersey State Department of Transportation as a Repairman at a salary of $6430 per year. I had no idea what my duties would be, but it was a job and I was determined to prove that I could take care of myself.

It turned out to be a job I was fairly well equipped to perform. I worked out of the maintenance shop at the Department of Transportation headquarters in the Trenton area. It was just down Parkway Avenue from my high school. I was essentially an apprentice to several tradesmen. I worked at different times as an assistant to carpenters, plumbers, electricians, painters and masons. Sometimes we worked there at the headquarters complex and sometimes we traveled the state repairing or remodeling transportation department buildings. In a few cases we actually constructed buildings such as salt bins.

My mental state fluctuated daily, sometimes hourly. Some days I would drive to work and lose myself in a particular task without the obsessive machinations taking over. At other times something would set me off and my body would instantly flood with liquid fear. An episode could keep me stuck for days in a downward spiral of negative thoughts.

One of the other Repairmen was a blond long haired young man named Jimmy. He was permanently assigned as the assistant to a middle aged plumber named Jude. Jude and Jimmy both lived in a town just up the river and rode together to and from work each day. One morning Jude was sitting on a bench in the shop area where we gathered to await our daily assignments. He had his head in his hands and was rubbing his eyes. Two of the other older men drew me aside and told me that Jude had stopped to pick up Jimmy earlier that morning and found out that Jimmy had killed himself during the night with a shotgun because his girlfriend had cheated on him.

I immediately thought back to the morning before when Jude and Jimmy had been sitting side by side on the same bench. In the early mornings the animated cast of characters that worked in our department would gather around the two benches outside our manager's office to await assignments, fortify themselves with coffee and swap banter and bullshit, especially deer hunting stories. I remembered that Jimmy had been quiet and very still the day before and soon convinced myself that I had had a premonition about his suicide. I began to obsess about why I hadn't paid more attention or said something. I also became frightened about whatever psychic mechanism allowed me that kind of insight. I was fighting very hard to make the world a concrete, finite and therefore manageable place. The notion that I could have read Jimmy's thoughts was exactly the kind of thing that undermined the world view that I so desperately sought to believe in. It also triggered other fears. I became afraid other people might be able to read my own thoughts and that was something I most certainly wanted to avoid. I was working very hard to conceal my disturbed state of mind and to fit in amongst people that were very different from me, even in my most stable moments. Jimmy's suicide again raised the fear that the world was truly an evil place from which there was no relief and no escape.

I began to think about the unbelievable number of people I had already seen try to escape the horror by suicide. I was only twenty-three but I had known far too many already. In Catholic school we were taught that suicides went to hell. That idea that someone who was in such pain could be sent to a worse place was impossible to reconcile with the idea of a loving God the nuns were selling simultaneously. It was frightening and painful to think of the people I had known who had died by their own hand, but for the next several days I was unable to stop myself.

There was Jimmy of course and then there was high school classmate Danny Lopez who stepped in front of a truck while on hallucinogenic drugs. My Uncle John ran a hose from his exhaust pipe to his car window and went to sleep forever in Cadwalder Park. The shop owner, whose son used to beat me up, had killed himself with a shotgun in his wife's bed supposedly because she wouldn't sleep with him and because of stock market losses. A girlfriend's father had hung himself as did a neighbor who went to the trouble to hang herself in my Aunt Eileen's garage to keep her own children from finding her. I had studied together with a girl at college who was being pursued by a kind but oafish boy named Ralph in whom she had no interest. Often as I dropped her at her dorm he would be sitting on the steps looking forlorn and would try to speak to her as she tripped up the stairs by him. One day as I dropped her off some of the other girls ran out to tell us Ralph had killed himself with a pistol to let her know he couldn't live without her. There were others. By the time I was in my mid twenties I had known twenty people who had killed themselves. I thought of Dr. Rittmayer's fears that her thoughts were causing harm and death. I worried that something similar could be happening to me. It was an easy scenario to get lost in but I continued to fight with all I had to ground myself and to stop those thoughts from taking me away.

After a time, I was assigned more or less permanently to the shop at the headquarters where I was taught rough cabinet making by the carpentry foreman. I worked for several months straight building workbenches, bookcases and other furniture to be distributed to the Department of Transportation sites around the state. I was working with a large radial arm saw and a table saw, a drill press, sander, lathe, a router and a jointer planer. I learned to order lumber and to schedule the cutting and assembly of the various components in such a way that lumber and finished projects would flow through the shop without getting in each other's way.

The work was relatively easy but required full attention around the power tools. That often helped me to keep my mind from running rampant toward my fears for too long. A stoner named Brad Edwards worked alongside me in the shop at times. He had long hair and all the speech affectations and mannerisms of a pot smoking hippie. He would sometimes go on about cosmic consciousness and various other hippie culture subjects. On occasion he could do a half an hour or more on why only vitamin C with rose hips was worth taking or why he could only eat poached eggs with cheese melted on top. He could do a whole afternoon on the relationship between peace and cosmic love and the reasons why he might kill someone before he let them send him off to war. He would do this at a sufficiently raised volume to be heard over the almost constant drone of the saws and other power tools and the hammering together of the various component of the furniture we were building.

The need for me to stay safely focused and attentive while working in the shop made Brad's occasional trips into the cosmos seem a bit less serious by comparison. This had the decidedly ameliorative effect of keeping me grounded in the moment and less likely to chase off on my own mental forays into the way out there.

Nonetheless, it was still an ongoing and constant struggle to believe the world was anything but a gossamer illusion. I was constantly on watch for signs that my physical environment might at any moment dissolve into something worse. If I misplaced something or if I saw something had changed without a plausible reason, I could spend many hours and even days in a heightened state of anxiety. I would search high and low looking for a misplaced flashlight or set of keys fearful that they might have dissolved into thin air. If something had been moved or been turned around I would set off on a Holmesian quest to determine whether natural or supernatural causes had been at the root of the change. If my investigation failed to produce a reason to explain something that had happened, I was completely unable to let the incident go. My emotional state could spiral downward and my fearful suspicions would then be projected onto everything and everyone with whom I came into contact.

Songs on the radio could taunt me. Signs on the highway could have hidden messages veiled to everyone but me, the single person in the entire world that message had been designed and created to communicate with at that very moment in that very place about the very thing that had just been going through my mind.

Even clouds in the sky could cause me to go into a spin. Standing outside the shop one windy day, I looked up to see a cloud dissolve. Now I had studied meteorology for a full semester at college and had received a perfect score on every test and exam, never missing a single question. It should have been child's play for me to conjure a simple explanation of how changing air pressure or temperature had caused the tiny droplets of water that made up the cloud to evaporate into invisible vapor. That explanation would not come to me. My anxiety level was already elevated due to some minor misperception or misunderstanding earlier in the day which had again cast doubt on the substantive nature of my world. Dissolving clouds were no less than a clear message that nothing was solid, that nothing could be counted on to be as it appeared or to stay the same.

For months I began to warily watch the clouds. As I looked more closely I found I could see images of faces and of bodies in the clouds. I could see animals and people and angels and demons. I could see distorted and grotesque combinations of all of these things. At times I recognized these as mere projections in the same way a child might see the twilight shadow of a tree on a wall as approximating the shape of an elephant. At other times when my mind was already over stimulated or churning with some earlier fear, the images in the clouds could appear as symbols of the deeper unseen undercurrents of the cosmos or as omens or warnings about events about to unfold in the path ahead of me. These fears could build upon themselves inside of me, causing me to project this increased level of anxiety onto everything with which I came into contact. I would then read additional unfounded meaning or symbolism onto nearly every situation. These painful spins would go on for days.

I had worked at the New Jersey Department of Transportation for about a year when the grounding effect of the daily physical work along with the weekly support from the therapy sessions with Doctor Rittmayer began to have a positive effect. I still did not trust the world or its intentions toward me but I was spending less time worrying and obsessing about it.

I was installing drawer slides in some workbenches as I watched the wall clock tick off the last few minutes of the workday. It had been a long and tedious day of mostly cutting lumber for the several benches that I was now hammering and screwing together. My ears were ringing from the hours of listening to the 16-inch radial arm saw and the 12-inch table saw chew through plywood and two by fours. Tony Napolini, one of the shop's two electricians, walked into the shop after a day on the road.

"Whatta youse doing?" Tony asked me.

Several of the Italian guys that worked out of our shop had chosen civil service jobs over higher paying work because they valued time with their families over status or more money. Most of them were relaxed, outgoing friendly people. Two of them, Chick and Dali, were brothers-in-law that couldn't do enough for other people. They were constantly teaching me things about the job but also about how to make meat sauce or to fry eggplant.

Tony Napolini on the other hand had a hard edge. He had a temper that was too close to the surface, the kind of vengeful mean streak that goes along with feeling like you've been passed over or never been properly valued.

"I'm making new workbenches for some of the maintenance yards in South Jersey," I told him.

"Nah, youse ain't makin' nuthin'," He spat at me. "Youse just think youse are makin' somethin!"

Jersey banter... fughedaboudit, right? Or tell him you don't have time to make nuthin' cause you been at his house makin' his wife all day. But that's not where my schizophrenic mind went. My mind made a giant and immediate leap backwards into fear. For a brief moment I saw Tony Napolini as an agent of the devil. He was telling me I wasn't making anything because ultimately nothing was real. He had been sent to remind me of the true black nothingness of eternity, of the pointlessness of the illusory shadow world we pretend is reality. He had been sent to pull the rug from beneath me so I would fall endlessly through the dark terror filled void that awaited me at the edge of my conscious mind and assuredly at the end of my life.

I froze. I could not answer back. I continued my work until four o'clock and quietly left the shop without a word to anyone. I did not return to work the next day or the next. In my next therapy session, I told the doctor I had decided to quit the job.

She helped me talk through the incident. When my mind leapt into fear a part of me knew almost as quickly that what I was thinking was ridiculous. Tony Napolini didn't have the presence of mind to be that duplicitous; mean yes, but that deceitful, not a chance. I knew almost immediately after it happened that I was projecting my fears onto Tony. Nonetheless once the fear hit it was almost impossible to stop for days and the fear and my negative thoughts energized each other the way the wind whips a prairie grass fire, then the rising heat from the fire pulls more air toward it.

I was determined to quit and Dr. Jane did not fight me about that. She may have believed it was time for me to move on. I had taken the job to ground myself and to prove I could take care of myself and as far as that went I had spent a year making progress in those areas. The doctor was insistent however that I return to the job site and give notice in person. She said she did not want me to look back later feeling that I had been beaten by my fears.

The next morning, I went into the shop and told them I was leaving permanently to continue my education, which was the direction I had settled on. I explained to the bosses exactly what had happened even though I knew they would not understand to any great extent. I ran into Tony Napolini and said hello and goodbye without more than slightly elevating my anxiety level.

I returned to taking more classes at Trenton State College. I was offered a part time job at Englishtown Raceway Park, a National Hot Rod Association sanctioned drag strip about an hour from home. I became the head of the stock classification department, in charge of assigning each car to the proper class according to its advertised weight and horsepower. As a small boy I used to pour over Motor's Auto Repair Manual and for most cars made in the fifties and sixties I was already pretty much a walking encyclopedia of which car came with which combinations of carburetion, cam timing, compression ratio and the like.

I had a crew of four including myself. One of other guys was nicknamed "Penguin" and like me, referred to the extensive manual we used to check serial numbers and other items to verify we had each car in the right class. Penguin and I each had a "marker" working with us to paint the number and class designation on the windows of the race cars using water based white shoe polish. In the afternoons I ran the staging lanes, communicating with the tower via a headset to insure cars from the various classes were in the correct lanes and that the proper pairings were brought to the starting line at the right times. When the head of the track was on vacation I was temporarily promoted to run the entire event from the top floor of the tower, scheduling each of the races throughout the day and calculating handicaps for elimination rounds in which class winners from earlier in the day ran off against the other winners from different classes. I manually calculated the appropriate head start to be given the slower car and then I set a time delay into the control box for the "Christmas Tree" lights at the starting line.

On Wednesday nights the track ran time trials and I was hired as the starter, staging each car at the starting line and operating the light tree that showed when each car was in place and then signaled the start. It was a noisy and smelly place to work. Screaming tires and open exhaust pipes competed with the sounds of crew members and drivers shouting last minute instructions. The loudspeaker carried the constant commentary from "Blaster," our excitable and sometimes stoned track announcer, Bobby Dorr. The odors of burning rubber, Clorox bleach, in which tires were spun to heat them up; raw gasoline, spilled oil and exhaust fumes added to the cacophony of sensory input that had to be filtered in order to stay focused. The occasional car coming out of the gate sideways and the not so rare fires, engine, clutch or transmission explosions at the line all kept my attention focused on my job to the exclusion of the crazy thoughts which could more easily take root in idle moments.

CHAPTER 19
I-295

I carried with me a deep sense of darkness that seemed to have worsened as I sat through the various Criminal Justice classes. Nights were blacker and there seemed to be a deep well of blackness within my body. It wasn't emptiness or nothingness but rather a fullness of pitch black energy. It felt like while my body existed in the atmosphere on the surface of the earth it simultaneously existed frozen inside a giant planet of solid coal. Images of guns massively increased the intensity of the feeling in me quite suddenly but the feeling was constantly and profoundly with me. At times the darkness would seem to creep into my peripheral vision as if it was leaking in from my brain.

At night I began to jolt awake from nightmares in which I was lost and desperately trying to find my way through very dark places. I saw angry and fearful faces everywhere I looked. Lying on the bed as the outside light faded, a myriad of wrathful faces stared down at me from the patterns within the popcorn ceiling. Faces of fear, anguish and aggression looked at me from the clouds, from folds in fabric, the outlines of trees in the woods and even from swirls in the potatoes on my plate.

I could understand the mechanism of how my brain could complete a partial or shadowy image with details that weren't there. I even understood that many normal people imagine things as complete without full information, sometimes to make us more comfortable. What I could not understand was the disturbing negativity of the images… and why these things were in my thoughts.

When I shut my eyes something much more extreme would happen. The faces of demons and monsters would rush up out of the blackness at me! These were far more striking than the shadowy ambiguous images that I saw in clouds or textures in the ceiling. They seemed "realer" than real and like they had lives of their own. I would jolt when they appeared. The visions were complete, sharply focused and vividly colorful. The colors stood out from a darker than black background the way florescent or day-glow colors jump out under black light. Devils heads, monstrous ogres and other frightening supernatural beasts would come at me one after the other. I learned that psychologists called these phenomena "eidetic images" and defined them as clear, lucid mental images that elicit an emotional response. They are described as being so much more vivid than ordinary mental images that they are often perceived as coming from an external source, outside the mind. Ancient Greeks believed these visons had healing powers and called them "gifts of the gods." Teachers of some Eastern religious disciplines describe similar experiences as visions of wrathful and peaceful deities that appear to spiritual seekers at certain phases of their development.

I had these powerful and invasive visions for several years. Sometimes I would see distorted surrealistic scenes that depicted torture or other atrocities. I had witnessed the torture of a young boy by the Vietnamese National Police and I had lived through years of a form of torture myself by way of the daily and brutal beatings I had suffered. I had also seen many forms of violence and the results of it during the war but I did not at this time connect any of the images that flooded my mind with my own experiences or even my own unconscious. Instead I was convinced that these horrible spirits where inhabitants of the void I was so afraid of being lost in. They increased my tendency to be constantly on guard, watching every external sign and every one of my own thoughts to try and avoid leaving open any access point for the spirits to invade my reality or for me to slip permanently into their world.

I began having to drive a fair distance at night once a week. Dr. Jane Rittmayer left her position as a staff psychologist in Dr. Jack Ward's office and was practicing on her own. She explained to me that her office was in Haddonfield, sixty miles from my parents' home and a good hour and fifteen minute drive. I had no choice but to make the drive. The only small measure of comfort or hope I had felt in the last year or so was through my sessions with Dr. Jane.

I had made a similar drive every working day for six months when I was seventeen and attended the General Motors Training Institute in Moorestown, N.J. where I earned my certification as an automobile mechanic. Those drives were made in daylight hours. At that time, six years earlier, part of the trip was made over the New Jersey Turnpike which was a toll road. Now a new toll free road, Interstate I-295, paralleled the Turnpike route from Bordentown, just below Trenton, to the Haddonfield exit, a few miles from Dr. Jane's office which was in a residential building in an older suburban neighborhood.

I-295 cut through uninhabited woodlands most of the way with tall stands of maple and oak on either side. There were no billboards yet and few homes or buildings on either side for many miles at a stretch. The new road was still very lightly traveled so on many nights I drove for miles without seeing any lights from other vehicles. I would sometimes become afraid that ahead in the blackness, the road was about to end. I was afraid there was a huge drop off ahead where not only the road but reality itself would end. Reality seemed such an illusion at times and my paranoia would convince me that the only thing out there was the part of the road ahead that was being illuminated by my headlamps.

The next thought that forced itself on me was whether I had the courage to test that theory by turning the lights off. Fear would build in me to levels that are surely unimaginable to most normal people, except possibly in their nightmares. To turn the headlights off was to risk finding myself in the eternal black void at the mercy of the demonic spirits I so feared. To not turn the lights out was to give in to that particular fear and effectively accept that belief as true. I would drive for many minutes hoping to see the lights of another vehicle while trying to build the courage to push in the headlight switch. My fear was that if I shut the lights off, when I turned them back on there would be nothing there. That's how tenuous my sense of reality was.

On some trips I would finally muster the courage to slam the switch shut for just a second then yank it out almost as quickly. It was truly pitch black with the lights off and the unanticipated fact that the dash lights went off at the same time exaggerated the effect. The relief I should have felt as the lights came back on, showing the roadway ahead unchanged, did not take hold. Instead I would instantly obsess that I hadn't left the lights off long enough for the illusion to have disappeared and that I must leave them off much longer to provide a true test. By the time I reached Dr. Rittmayer's office I was sometimes physically exhausted and so emotionally strung out that my mind would go numb.

As I sat in the small waiting room outside the doctor's office I would forget many of the issues that had arisen during the week. As I stared at the florescent sheen that bounced off of the wood veneer paneled walls, all I could think about was whether my final encounter with the black void and the hell within it would happen on the drive home that night.

CHAPTER 20
Green Orchids

In June of 1972 I had been suffering with schizophrenia for two full years. I had periods of relative stability during which I was able to work or to go to school. I was faithful to my high protein, low carbohydrate diet and I abstained from alcohol, caffeine, illegal drugs and for the most part from sugar or anything else that Dr. Ward said could destabilize my brain chemistry. I took the massive doses of megavitamins that were prescribed as well as the antipsychotic medications and the other prescriptions that controlled my blood sugar.

I had reached a point where my daily torment was not as frequent or intense but I did not yet feel anything close to right. The fears and the mental obsessions were more under control but they were still very much there, unresolved in the background. I felt dead inside and numb on the outside. Occasionally some small flash of emotion would break through to the surface.

I had been out to dinner with the same girl several times when she asked me to accompany her to a dance. I knew a woman who worked out of her home as a florist. She had lost a hand somewhere during her life and everyone was always justifiably impressed at what she could accomplish with floral arrangements. I called her and told her the color of the semi formal gown of my date. The florist suggested green cymbidium orchids for the corsage. During a phone call with my date, on the day of the dance, I happened to mention the color of the corsage. Apparently she didn't like the flower choice by the florist and she started yelling at me over the phone. Now I had been through quite a few things in my life and I didn't think this was the kind of thing anyone needed to yell about and I told her so. That only increased her anger. By the time I hung up the phone to call the florist back I wasn't real sure I wanted to go to the dance.

I sat down to gather myself to call the florist at this late moment. As I got back up to go to the phone I was overcome by a wave of tiredness. I sat down again to get my bearings and realized I was burning up with fever. I went into the bathroom and wet my face to try and cool it down. At that point I was under the impression that the unexpected emotion from the phone call had me off balance. During my illness just about anything could knock me off my tracks. Then I tried to urinate and could only pass a few drops of brownish syrup.

I was taken into the emergency room where I was met once again by Dr. Jack Ward who still had sway over the medical aspects of my case in spite of the fact I saw Dr. Rittmayer for psychotherapy. I was admitted on the spot and told I was critically ill with hepatitis. When I reached my room I took time out to call my date and tell her I would happily not be bringing her green cymbidium orchids that evening… or anything else. I also called the florist and I have to say that she took the news much better than my date.

The inflammation of my liver was not the infectious type. It was a side effect from the medication Quide, the antipsychotic drug that had replaced Mellarill, which had replaced Thorazine. I had taken three antipsychotic medicines. All three caused severe side effects. Of the three hospitalizations during my illness, two of them, the first and third were directly related to reactions to the drugs I was prescribed.

Dr. Ward took me off the Quide and put me on a different class of drug. The three that had caused the problems were considered antipsychotics or major tranquilizers. The new drug, Vistaril, was a minor tranquilizer not often used to control psychotic symptoms, but rather anxiety caused by normal neuroses or stress from ordinary life events.

I stayed in Mercer Hospital for two weeks. I was desperately in need of the rest. It also gave me some time to reflect on my progress. The idea that I had been able to deal with my dance date's anger in the way that I had was a sign of positive movement. Just a few months earlier I believe I would have decided the girl was a witch out to damage me psychically and I would have cut off from her and run away from the situation. This time I was able to stay with my experience of her anger and to respond to it more appropriately.

Not long before when my co-worker had made a mean comment to me, I had leapt to the conclusion that he was an instrument of some evil presence that was out to undermine my grip on reality. I had left the situation rather than stay with the experience of fear long enough to see that my fear was merely covering up my own anger at his insensitive remark. When my date started yelling on the phone my first response was to flee into fear. "What had I done wrong to cause this?" "Oh, hell, I messed up again." I did briefly entertain the idea that the girl was some kind of witch that had drawn me into the situation on purpose. When I had the problem with the co-worker at my last job, Dr. Rittmayer had insisted I go back and confront the situation and I learned from that. When the impulse came to hang up the phone, I stayed on and listened, not only to the yelling but to my own feelings and I discovered that her response made me angry also. I was beginning to see that fear was not always fear. Sometimes fear was a way to cover up other less explainable or acceptable feelings. I didn't want to be angry. Angry people were bad or so I had been taught from a very young age. The anger hadn't come through strongly. I had only a hint of it, a hint as it were, of things to come.

On the new drug another level of physical sensation began to return. The Quide had less of a deadening effect on me than the previous antipsychotic drug but my senses had still been dulled and I was largely numb to my emotions compared to what began to emerge when I was switched to the much milder Vistaril. Bodily feelings slowly returned and as they did I found myself less concerned about whether I would dissolve into space. My perceptual senses became stronger again and the world seemed less distant and unreal. The flashes of emotion came through more frequently. At first the feelings were relatively subtle but, compared to having felt dead to anything but fear for two years, they definitely got my attention.

CHAPTER 21
The Dream

I sat numbly in front of Doctor Jane's desk half heartedly recounting the week's disturbances when my mouth suddenly got very dry. Whatever I was about to say slipped from my mind completely. The same feeling had come up before in therapy. The doctor had said something to me, which I was unable to follow and my thoughts had blanked out in exactly the same way they just had. She had commented at the time, "Oh, it's probably just something you're not ready to hear yet," and she had gone on to ask a different question.

This time, just as before, I felt as if some subtle electrical force had just drained all the mental energy from my brain. Not only was I unable to remember what I was just about to say but I was also unable to generate a single new thought of any kind. Confusion was the overwhelming feeling. I sat for a moment trying to grasp what had just happened. I stuttered. "I…I … uh I can't…"

The doctor folded her hands. "It's all right," she said. "If it's important it will come back to you when you are ready to talk about it."

After continuing to try unsuccessfully for a few more minutes I gave up and my mind began to clear. I went on talking about my usual fears of dissolving into the void and about some of things that had triggered my fear attacks during the week.

The next week I began to talk again much in the same vein when about halfway through the session I remembered that it was a recurring dream I had wanted to talk about the week before when my mind went blank. At first I could sense only the feeling of terror that I awoke with as I was jolted out of the nightmare with a convulsive spasm of my entire body. My heart had been pounding in my chest and I had been gasping for breath. I had looked around my darkened bedroom for signs of spirits or other souls in the room.

"I had a… a dream!" I choked out the words and as I did my body jolted sideways in the chair like I was ducking for cover. I saw a flash in my minds eye and a small piece of the dream came to me. I began to describe it to the doctor but as I did I began to choke up and tears rolled from my eyes. "They were… They were…" I tried to talk but now the images from the dream came flooding back and I sat motionless while I was overwhelmed by the pictures that flashed though my mind. It was a dream I had previously tried to remember enough to talk to Dr. Ward about but after he more or less dismissed it as unimportant relative to my diet and medication, I had forgotten it until now. When it came up in Doctor Ward's office, the images from the dream had been sketchy and incomplete. Now in Dr. Rittmayer's office they hit me with full force. I gasped for breath and my heart raced like a runaway freight train.

I was in a sandbag bunker at night that was under attack by enemy ground forces. Instead of the black pajama clad, lean and wiry Viet Cong that I had faced in Viet Nam the invaders in my dream were sturdy square faced soldiers with muscular bodies. They looked more like Koreans or Mongols than the thin and slight South Vietnamese. They were clad not in black but in pure white ghee's or karate outfits that gave off a luminescent purple glow which spread over the scene like ultraviolet black light. The soldiers advanced toward my bunker from a rice paddy next to the road, firing at us and throwing hand grenades as they came. There were more than a dozen of them and I was alone with one other American soldier. My rifle was jammed and I frantically worked at clearing it as the enemy came closer.

Finally, I withdrew a stuck round from the rifle's chamber and began to fire. I reached a superhuman level of performance directing rounds through the darkness with uncanny accuracy. I could clearly see the faces and even the eyes of the enemy in the eerie violet light from their white uniforms. I could see my rounds striking their bodies but still they came at us like machines. The bullets passed through their white ghee's and through their bodies without any effect. The cloth of the uniforms was unbroken and unmarked. The enemy soldiers came on in slow motion unfazed by our rounds that had gone through them. Like phantoms they passed right though the barbed concertina wire that was coiled in rows in front of our position. They were somehow beyond the effects of our firearms or other defenses as if they were operating in another dimension. They continued on advancing forward, gliding like ghosts, at an even unbroken pace, through the barbed wire then through the double sandbag wall of the bunker and finally through my body itself as if I wasn't there at all!

I jolted and my heart pounded. I remembered that this was a dream I had had many times but was unable until now to remember much of it except the feeling of terror I awoke with. Now as I was shaken back from the dream into the chair in the therapist's office… I began to sob. I don't have any memories of crying as a child and certainly not in my adolescent or early adult years. My eyes filled with water, I gasped for breath… and I began to cry. I tried to hold it back and it got worse. I sobbed and wailed and the tears and my saliva soaked the front of my shirt. Several times I tried to sit up and stop but each time the sobbing only increased. I cried from deep inside for many, many minutes. After a time, Dr. Jane got up from behind her desk. She walked around beside my chair, knelt beside me and gently put her arms around me. I continued to cry loudly and wildly for many minutes. I may have cried for twenty minutes before I was able to gain any control.

I was embarrassed but even more I was bewildered. It was a completely new experience. I felt warmth in my chest and belly where there was usually icy cold fear. My arms and legs were in some small measure relaxed a bit from the usual knots they were tied in.

The doctor soon went back behind her desk. "It's a good thing, Tom. This is a very good sign." She told me to go home and rest and that we would talk more about what had happened soon.

The dream paralleled an incident I had experienced in the war. I had been alone in a bunker with one other Military Policeman. The bunker was on the perimeter gate of a forward base camp that had just been taken over by the 173rd Airborne Brigade. We were at the edge of a village of unarmed Vietnamese civilians and our bunker was the only one with a view to the front of the village. A force of thirty Viet Cong came up out of the rice paddy and began throwing grenades and firing automatic weapons into the village, wounding more than a dozen unarmed villagers and killing a six month old infant.

My partner and I were the only ones close enough to respond and we fired on the Viet Cong, drawing rifle, machine gun and grenade fire onto ourselves. My rifle had actually jammed in combat as it did in the dream and after I frantically cleared the malfunction in real life, I seemed to be overtaken by some extraordinary clarity which allowed me to achieve some unexplainable level of performance as we fired on the enemy. Unlike the dream in which our bullets had no effect, we killed at least eight and wounded an additional unknown number according to the follow up investigation by the local military intelligence unit. Several of those that I killed, I saw fall before me close enough to have seen their faces as they fell. My partner and I both received high praise from the villagers, from our commanders and from their Vietnamese army counterparts. We were told we were to be put in for decorations for valor by both the Americans and the Vietnamese, neither of which we ever received.

I had no real understanding of why I cried when I remembered the dream. It would be years before I would fully comprehend the intensity of my feelings associated with that particular war incident as well as several others. In the meantime, the fact that I had felt something besides blinding fear and anxiety was encouraging to my therapist. I did not yet understand the importance of this breakthrough but the doctor's enthusiasm helped me to believe something crucial was happening and I continued to work at trying to remember my dreams and to talk about them.

CHAPTER 22
Notes to My Self

When the memory of the firefight dream came back to me in Dr. Rittmayer's office it was the first time since the beginning of the breakdown that I had been able to fully remember a dream. I often jolted awake from night terrors, shaking and with heart palpitations but I hadn't any real sense of the dream content. At other times the memories were partial and the feelings that had awakened me were lost before I was fully awake. Now the surreal details of the combat nightmare came back vividly again and again for many weeks. Always I would awaken at the moment the white clad phantom warriors with the purple glow would pass through my body as they charged forward.

Soon other dreams began to surface. I had a recurring dream of trying to cross a bridge high above raging currents. Always I was clinging desperately to some flimsy part of an otherwise massive superstructure, focused intently on trying to get to the other side. Some part of the bridge was always incomplete or deteriorating. It was dark and I was high above the swirling currents and the wind howled around me.

I had another dream several times that was a variation on a nightmare I'd had many, many times in childhood. I was trapped within the walls of a house that was similar to the home I lived in during most of my grammar school and high school years. I would enter into a kind of secret storage area under the eaves and become entangled within narrow channels and passageways through the studs or skeleton of the house. I would become frantic for breath like I was being held under water. I could feel that the outside of the house was close by but I could not find my way there.

With Doctor Jane's insight and support I was eventually able to understand that the house represented me and that I had some how gone inside myself to a secret place that seemed safe, but became trapped there. To begin to understand this dream and to experience fully the feelings from the dream while I was awake was among the bigger turning points I had in the healing process although I can only see that in retrospect.

To be able to experience my fear at that level of intensity while awake and conscious was a major breakthrough. I had lived for way too long being afraid to be afraid. Now I began to be able to experience increased levels of fear for longer moments and to my great surprise I did not dissolve into the void nor did I drop through the floor or become absorbed into some new hell. This was an epiphany on many levels. The idea that the source of my fear could be my own unconscious thoughts, was shocking.

Dr. Jane gave me a paperback book by Hugh Prather, called "Notes to Myself." She said although she understood that my fear and confusion about the supernatural was my biggest concern, there were other areas to be addressed. The book was a kind of poetic dialogue between a man and his inner self. It seemed so simple and easy compared to the battles I had been fighting that I couldn't really see the relevance initially. Eventually I began to understand that I had no contact with my inner self except through the glimpses I was beginning to get through the nightmares. The things that had made me afraid for so long, I had seen as outside myself; so far outside myself and so removed from me that I believed they were supernatural.

I slowly began to find the strength to look inside at my feelings, using the access points provided by the nightmares which became more frequent. I worked at trying to remember them more fully and developed a kind of technique that helped me to do so. When I awoke, rather than trying to remember the specific details of the dreams, I would try to stay with the predominant feeling from the dream. In the beginning that was always some form of fear. I disciplined myself to start my day slowly, trying to touch back into the primary feeling from the dreams rather than losing myself in some activity right away. Often at some point during the day while I was trying to focus on the fear from the dream, some mental image or thought would surface which helped connect the dots to complete the story told by the dream. As my ability to handle feelings from the dreams improved I found that the material from my dreams would often relate to unresolved feelings from my life situations, sometimes from many years ago. When recalled and faced fully, sometimes in my therapy sessions and sometimes sitting alone, the fear would finally subside. I was left with a feeling of increased inner strength and a kind of clarity not unlike the clear headedness that invigorates people after a thunderstorm. When I was able to stay with or penetrate the feelings, the energy from the negative feelings would be transformed or more accurately transmuted into some positive strength. This was a giant step forward in my recovery.

I renewed my efforts at school, taking mostly day classes again. One day when I was on campus I found a note on the bulletin board advertising a part time job as a woodshop instructor at the inner city after school and evening day care program run by the Boys Club of Trenton, which in spite of it's outdated name had for some years provided services for both boys and girls.

I drove downtown and was hired on the spot. I began to work part time guiding inner city kids through small projects with hand tools. The projects were designed to instill confidence in the children's abilities and had the side effect of creating needed dialogue at home as many of the small projects were made to be presented as gifts to family members.

My situation improved sufficiently for me to move out on my own. I left my parents home in Lawrenceville, NJ and took a small apartment across the Delaware River in Morrisville, Pennsylvania. I would occasionally take a couple of pieces of wood, metal or plastic home with me from the shop and play around with them trying to come up with new projects that were simple enough for the boys and girls at the club to do.

I had a beat up old black and white television that relied on a jury rigged aluminum foil antenna for what limited reception it got. The sound wasn't working on the set so I got another old small set someone had left on the curb which had no picture but did have sound and between the two sets I was able to get some manner of full reception. The furnishings in my apartment were scarce. I had a twin bed from my parent's home, a couch that was discarded by an aunt and a couple of other odd pieces of outdoor furniture I had scavenged from other folks' trash during the annual "Clean Up, Fix Up, Paint Up" week.

One night I was holding a ten-inch-long block of walnut hardwood that was a couple inches square on its shorter sides in my hand as an old detective movie was coming on the tube. As I became engrossed in the story I began to carve away somewhat unconsciously at the block of wood with an Exacto knife and a couple of other tools. One movie ended and another began. When the second movie was over I found myself holding a hand carved walnut bud vase. Overall it had a subtle hourglass shape. The angled opening at the top was curved and flowing. It had elements that kind of combined male and female sensuality or eroticism. It was beautiful.

I was stunned. I did not know the person within me that had carved this beautiful object. I was staggered to think that there was a part of me with this ability that was entirely outside my conscious awareness. I began to wonder if I could draw.

CHAPTER 23

The Artist in Me

I bought a large sketch pad and two charcoal pencils at the discount mart. The next night I again turned on the two televisions to the late night movie channel. I set pencil to paper with out any plan at all. I drew a curved line and then another. As it had been the night before, some large portion of my attention became absorbed in the unfolding story in the old movie. I would look up at the screen long enough to register the faces of new characters as they walked on or to take in a new location that the story had moved to. When I looked down at the drawing pad, my mind was still following along closely with the movie story. Although I had voluntary control of my hands, it is more accurate to say that I was watching myself draw than to say that I had any real cognitive input into what showed up on paper. After all, I had no idea how to draw, had never drawn before and if I'd had to explain to my hands what to do in any way, I would have been at a complete loss.

The first drawing was a surprisingly well articulated hybrid mix between a panther and a snarling mastiff of some kind. If pressed to call it something I would have titled it "Hellhound." It did not have a finished look as if it had been done by a professional but it was a far cry from primitive. Above all it evoked a specific and powerful emotional response. The drawing seethed with ferocious anger.

I was again more than surprised that this drawing had come from me. I was bewildered by my ability to complete such a drawing and I was just as surprised at the intensity of the emotional content.

I continued to draw through the night turning out increasingly abstract images that portrayed confused family hierarchies, repressed sexuality and loneliness. The message or feeling conveyed in most of the drawings was not at all clear to me as I drew. My conscious mind was almost completely absorbed in the TV movies as I worked through the night. It was not until I looked at some of the drawings the next day that I could clearly see the content or emotional impact of some of the more abstract pieces.

I showed the bud vase and the drawings, to Dr. Rittmayer. She practically leapt out of her chair to inspect the Hellhound drawing as I held it up. She looked at each drawing several times as she encouraged me to continue drawing. It was clear to me that she saw this as a way to access my inner workings. She tried to get me talking about the Hellhound. I was aware that she was interested in getting me to investigate the anger expressed in the drawing. She could see that this was another major breakthrough to a new level of connection to my real self and to emotions that had been covered up by the terror attacks, because I was literally too afraid to experience the hidden feelings.

I began to experiment with other mediums. I purchased pastels, watercolors and eventually acrylic paints. I tried working on different kinds of artist paper, and on wood, Masonite, canvas and even sandpaper. I spent many nights with a drop cloth spread over my apartment floor, painting or drawing until early morning while my mind followed the story line from a black and white forties movie or as I listened to an outdated record I had gotten from the discount bins at the Neshaminy Mall.

I turned out about four or five hundred drawings and paintings in the course of about a year as well as filling a couple of sketch books. I would fall into a kind of trance or meditative state for hours at a time, always late at night. My body would engage in the artwork and my mind would go off on some adventure somewhere with Humphrey Bogart or James Cagney. Somehow this allowed powerful and long withheld feelings to surface in me and be recognized and accepted as my own.

I would show the week's works to Jane and they would provide a jumping off point to talk about what I had felt. This led to discussions of where those feelings had come from. I began to open up and my therapy sessions were expanded from one hour a week to two hours once a week.

At times during the therapy I would find myself overwhelmed by the same wave of blank confusion that had preceded the memory of the dream about the firefight. I began to recognize this as a defense mechanism against the overpowering feelings that often accompanied a breakthrough. Because the experience had been so freeing in the long run, I began to work hard to keep going over the same material and trying to bead in on what had just preceded the moment when I had gone blank. It was hard work. The confusion and blank mind were often accompanied by a feeling that resembled nausea. It felt as if my entire body wanted to revolt or throw something off.

I began to watch for these signals in situations outside of therapy as well. Sometimes material that came up in painting or in therapy would trigger dreams a few nights later. Sometimes the residual feelings from nightmares the night before or even several days before could provide a starting point for a painting. Clearly the firefight dream had opened a channel between what I had accepted previously as my "self" and some concealed part of me that contained hidden feelings and other material that was completely new to me.

I began to have many terrifying dreams on a regular basis. I was pursued by terrible demons that were out to devour me. I was almost always carrying a load of baggage that was way too heavy and unwieldy. I was always missing something from my belongings and feverishly trying to find and gather up the missing items. I was anxious and struggling to make some travel connection I was just about to miss. I was stuck inside of dark mazes or structures that were deteriorating or broken so that there was a constant possibility of being injured or falling into an abyss. The demons were ever present, just around each corner, behind every wall, just about to catch up with me.

Some of the dreams brought to mind images from the old black and white episodes of space traveler Flash Gordon. In the old TV serial there were clay men that would materialize out of the earthen walls in the underground caverns and pursue Flash like the demons coming after me in my dreams. The evil Emperor Ming had a dark energy about him that was similar to the evil in the void that I so feared. Ming put helmets on his prisoners that would blank their minds like the confusion that would hit me from time to time.

It was sometime in 1972 when these breakthroughs began to happen. For two years since the major psychotic break in the summer of 1970 I had felt nothing but terror, high anxiety and a knotted, tired and almost dead body. Now other feelings began to surface. The first real change had been when I cried as I remembered the firefight dream. Then the snarling hellhound had showed up in my first drawing. Initially I experienced the anger on its face as something external to me. Soon with the help of my therapist I began to slowly understand that the anger had come from inside of me. Before long I began to experience feelings of anger inside myself, at first in dreams, then while I painted and eventually in my fully awake or conscious state.

It was a process. At first I was afraid of the feelings and tried to reject them. As they became stronger and I became able to experience them without fleeing into fear, I was able to inspect them more closely and see how they could be the result of unfair or difficult situations that I had long forgotten. With Dr. Jane's help I slowly started to understand how those hidden feelings could have built up upon each other for so many years that the underlying force behind them seemed overwhelming or even unreal. As I began to uncover more and more memories of situations in which I could justify anger it helped to explain the frightening level of anger that was apparently in me.

Anger was the first of several levels of feelings from which I had long been separated. It was still early in the process of rediscovering my true self but already it was becoming clear to me how my mind could have projected horrible things to explain my fear rather than let me feel the simple but powerful feelings which I found unacceptable.

CHAPTER 24
Awakening from the Nightmare

Earlier, when feelings had tried to surface during the worst of my schizophrenia, my mind had reacted by creating fantasies or scenarios that distracted me from experiencing the depth of my discomfort. Now these things that I had trouble handling began to show up more forcefully when I was dreaming or when the reactive part of my mind was distracted by painting or drawing.

Initially I had only a vague sense of the feelings from my dreams but no memory of the images or situations that had triggered them. As I became able to stay with the feelings by accessing the meditative or hypnotic state I discovered while painting, the content of the dreams became clearer and my ability to penetrate the fear and to handle it grew significantly over time.

I have heard it said that it is not possible to dream that we have died or we would actually die. This was not true for me. During this period, I had many nightmares in which I was killed. I had one dream in which my Volkswagen slammed into some invisible and immovable barrier. I could feel the metal crushing and sense the glass shattering and the next thing I knew was looking out to the darkened roadway from the hood of the car. There was a mass of something on the roadway before me. Focusing closer I could see a pile of fleshy, bloody pulp. When I gathered myself to move I found I had no feeling in my limbs. I tried to blink my eyes and realized I wasn't looking through my eyes but through the top of my head which lay lifeless on the hood of the

VW. The mass before me on the road was my own brains which had spilled out in the collision.

I had countless such dreams in which my body died and my consciousness was still active. My biggest fear just weeks before had been that my body would dissolve and my disembodied soul would be lost forever. Now in my dreams I began to have some version of that experience and to be able to stay with the feelings of extreme fear that came with it. As those dreams went on over several months I slowly began to be able to stay with the experience of death longer and longer and eventually I was able to surrender totally to the fear. When I did I would sometimes feel that I was leaving my body and flying. Other times I would seem to stay in the same place but my body and everything around me would transform. The experiences were extremely powerful. The sense of reality was in some ways stronger than my waking reality. The external forms in the dreams were fuzzy or ill defined sometimes but the emotional feelings connected with those states were many times more powerful than anything I had experienced while awake, although feelings during certain frantic moments during the war were close.

Rather than awaken from the nightmares suddenly in a panic, as I usually did, I now sometimes found myself in a kind of dual awareness. A part of my mind was fully engaged in handling the images or situations in the dreams and the powerful feelings of fear that came with them but on some nights another peripheral part of me was simultaneously aware of being in my bed in my body asleep. On a few occasions I was able to shift my consciousness from the "dream body" to my sleeping body in the bed and gently awake myself. Much later I found this phenomenon described in works on the Yogas of Dream and Sleep by Eastern spiritual teachers. They talked about it taking as much as thirty years of meditative practice to achieve this state.

Dreaming or awake, I began to develop an ability to stay and fight through whatever feelings I was having. From this point forward I began to believe full recovery was possible. I still had my concerns about the nature of reality and the fate of my soul but now I had something tangible and workable to battle rather than the vague sense of the supernatural unknown. It took every ounce of courage I had to stay with the images and feelings that were raised by the dreams. I found myself drawing on the memory of having to stay and fight in impossible situations as I had to do during the Vietnam War.

In another dream, vicious spider like creatures were expelled from my anus then grew in size and turned to attack me. It was a dream that repeated regularly and for many nights I awoke in fear and disgust. In a session with Dr. Rittmayer, I told her of the dream and how revolting and frightening it was. She suggested that all the content in our dreams comes from some part of our own selves and that often each aspect of the dream represents some aspect of our self. Both of these were new ideas to me and took me some time to appreciate. I became determined to stay and look directly at the feces monsters rather than flee in fear. Soon the same nightmare returned and after a couple of attempts I was in fact able to plant myself firmly in place within the dream and look directly at the horrible spider creatures that were dripping with excrement. As they came forward to devour me I refused to move although I was bathed in fear. A remarkable thing happened. The monsters dissolved. They simply ceased to exist.

The dream about seeing my brains on the roadway came back. This time, instead of blacking out at the moment I recognized I was dead, I forced myself to stay in the dream. Again, waves of fear washed over me but I stayed with it and found myself in a half awake half asleep state in my own bed. I was still dreaming but at the same time I could tell I was in my bed. My body literally convulsed as the fear washed up through me. As the feelings subsided or played themselves out, I found myself fully awake in the bed and with a full memory of the entire dream, but no residual fear. Now my conscious self had truly become open to a channel to my hidden inner self.

Additionally, going through the imagined deaths and terror filled situations had left me with a sense of self that in some way seemed indestructible or at least far less fragile. I felt that I could draw strength or even a palpable energy from my fear rather than run from it. This was the beginning of the end of my disease. It was also the beginning of my own experience of myself as a spiritual being but it would be some time in the future before I could understand or accept that idea. For the moment it was more than enough to know that my terror had tangible causes and that world might be a workable place after all.

I was also becoming aware that I had no real sense of who I was. My ideas about who I was up to this point had been almost entirely situational. I had been raised in a way that made me stifle whoever I really was inside and adapt to the overly controlled home life that suited my parents. I had stumbled through school life unable to fully comprehend or relate to the narrow religious science and structured studies that had been preached there. In the military there was even more structure and in some ways it made even less sense. Only in actual field combat or other desperate situations had

I ever thrown out the rules and functioned on my inner intuitions. It was interesting to me that the feeling of inner strength that pulled me through the terror filled dreams, also recalled specific memories of having to stand and fight during the war.

CHAPTER 25
Daily Life

As a youngster one of the first television shows I ever saw was "I Led Three Lives." It was about an undercover FBI informant named Herbert Philbrick, who infiltrated the communist party. He lived one life as a communist party agent, a second one as an FBI mole and a third as a bank executive and family man whose neighbors had no idea at all about the first two scenarios.

I felt a bit like Philbrick. A third identity was emerging; a partially formed personality that believed I could heal. That part shared mental and emotional space with the terrified schizophrenic and with the tight lipped knotted person who did his best to conceal the broken part of me while trying to function at work or school. I was promoted from part-time woodshop instructor to the full time Social Education Director at the after school and evening day care program at the Boys Club in downtown Trenton. The woman who had held the job before me had a master's degree in social work, a requirement which was waived when she left and I was offered her job. I was as surprised as anyone at the offer. While I still felt like damaged goods, apparently my supervisors saw something else in me. I became responsible for all operations except the gym. I planned and scheduled activities for the woodshop, game room, arts and crafts, library, remedial tutoring, movies, field trips, and meal times. I hired and supervised part time employees for all those areas and arranged transportation to and from home for those children who needed it.

I was proud of my position and I understood the value in the work. Many of the families that sent their children to our government funded after school and evening day care program fully appreciated what we did. For some single parents it was the only way they could afford to have their children cared for from three in the afternoon until nine at night so the parent could work. Most also understood that the immersion the kids got in the club's recreational, educational and cultural pursuits was far better than whatever else they would be exposed to on the streets of downtown Trenton.

Discipline was easy in most cases as most of the children also valued what they had at the club. A simple suggestion that they might miss an activity by being sent to "the bench" was usually all it took to keep everyone in line. In the rare case when that was not enough, a suggestion from the Unit Program Director that the next step would be to call the child's parents was almost always enough to calm down the most street tough defiant kids in short order. In one unusual and tragic incident a twelve-year-old boy's parent was called and he was picked up and taken home. The next day we learned he had committed suicide by hanging himself from a doorknob with a belt. On top of the normal impact of such a tragedy, the event sent me back into a brief downslide as I began again to obsess about evil in the world and about the devil waiting at the door of death to torment us further.

My commitment to the job was complete so far as my duties to the children and my supervisors went. I started there as a part- time employee earning just enough to make ends meet while I attended school. Now I had a real job which had given me a certain amount of self respect but I had no sense that I would continue in the job forever nor did I know yet what I wanted to do instead.

Doctor Rittmayer suggested that I might have sufficient talent to pursue art as a career. She asked me whether I had any interest. For quite some time and possibly for my whole life to that point, I had had no concrete direction and no real plans other than to try and escape whatever uncomfortable situation I was living in whether that was beatings as a child, the Vietnam War or my bouts with terror and depression.

I had nothing to lose by trying, the doctor pointed out and she arranged for me to meet with a successful artist she knew. He invited me to visit him at a couple of mall art shows to understand how most of his contemporaries made their livings. Without exception each of the artists had developed a "schtick" and continued to churn out variations of the theme that sold successfully for them. Some of them did nothing but kitty cats; some did dunes and seagulls, and so on. None of this did anything to increase any tendency in me to identify with being an artist.

Nonetheless, my fascination with what was coming out of me and the improvement in my condition that came from using the paintings to generate material in therapy, kept me working hard. At the encouragement of my old high school friend David, who was a professional portrait artist, I entered three paintings in a juried art show. To my great surprise two of my paintings were accepted while all three of my friend's works were rejected. A short time later I was offered an opportunity to exhibit my work in a month long, one man show at the Mercer County Library. Not long after that I had another month long show at a country club.

I took a film course at Trenton State as an elective. I did not do well in the course as my shyness and insecurities prevented me from enlisting needed help from other students to complete required projects. Nonetheless I believed I had found my passion. My painting and drawing skills were primitive. I had no formal training of any kind. All I could do was experiment and although I had some success with certain mediums, I felt limited. Filmmaking was a different story. The tools and elements of film were something I could immediately grasp and control. It was a whole new language that had the potential to express the way I saw and felt about things. The seed was planted. I knew what I wanted to do.

I continued to work at the day care program, to attend minimal classes at college and most of all to paint and draw into the night. With each new work I discovered something more about what was hidden deep inside me and about who I really was. I carried the paintings with me to my weekly two-hour therapy sessions. Dr. Jane and I went over each one reviewing insights or feelings that had come over me when I painted or when I had looked at the works the next day. During the therapy, investigation of those feelings and insights sometimes led to major breakthroughs. A feeling associated with a drawing could connect to the memory of a moment in my life where I disconnected from the way I felt in order to conform or behave in a way acceptable to someone else. That could then open up an entire area in which I had for years behaved against my best interests by continually suppressing my true feelings or perceptions. One of the areas uncovered was about my feelings and behavior toward girls.

CHAPTER 26
Girls

One night I experienced strong feelings of sadness and regret as I painted an abstract watercolor dominated by dark shades of blues and purples. The sky in the painting was filled with images of sad faces that seemed to be dissolving and raining or dripping down onto a stone bridge that spanned a dark and swirling river. Looking at the painting in therapy, that sense of sadness intensified into an extreme feeling of loss. I kept thinking of my race car partner's wife, Dot. I had always had a special friendship with her but I had never let myself think past that.

During that session I was able to see that my feelings for Dot ran far deeper than I had been able to accept consciously. Jane helped me explore the idea that having feelings are not the same as acting on them. Rather than feeling guilty about being attracted to my partner's wife, I could feel proud of the fact that I was a moral enough person to have kept my feelings in check. Looking further at my history with girls I recognized that there were quite a number of girls who I had been drawn to but I had usually worked to create friendships with them, truly afraid to cross the line into dating or courting. It was another example of how I had walled off from my true feelings in order to behave in a way I believed people wanted me to.

I did not receive affection at home as a child. I don't remember being hugged. Even as a toddler in Trenton my mother would walk me on a leash rather than hold my hand. Only once in the house on Latona Avenue do I remember my mother holding me on her lap and it is not a pleasant memory. Looking back, I am certain she had been drinking. She had her arms around me and was rubbing her cheek against me. I felt trapped and very uncomfortable. She whispered to me that a mother always loves her firstborn more. Even at my young age of eight or nine I knew this was very wrong to say to a child. She had saddled me with a sick secret that I would have to hold inside forever to keep from hurting my sister and until this very day I have never spoken of it to anyone. If I had needed physical affection I certainly didn't want it from my mother from that day forward. It felt sick the way she was holding me and talking to me. I felt violated and I was certain she was wrong to say she loved me more than my sister.

While I have only a few memories of feelings of fear or anger as a child I do very much remember strong and almost constant feelings for girls. Whether I was simply starved for affection in my home life or whether I had an unusually early mating drive is unclear. By the time I was in first grade in Seattle, I had noticed girls. Noticed is not a strong enough, I was in love with girls or more accurately with one girl, or again even more accurately with one girl at a time. You might say that I had the makings of a six-year-old serial monogamist.

I sat directly behind Karin Sundberg in Miss Magda Foss's first grade class at Coe School in Seattle. Miss Foss had white hair. Karin's was blonde. Magda's eyes were small and dark. Karin Sundberg's were wide and blue and darted shyly down to the side while her lips curled up in an embarrassed half smile.

I can vaguely remember printing in big block letters with an oversized pencil on sheets of wide lined yellow paper but I vividly remember the exact shape of Karin's cheek when I would glimpse it as she turned momentarily to the side. I can picture a few images of Spot and the picket fences from the Dick and Jane books we read that year but I can tell you the exact curve the curl of Karin's hair took as it swept down gently from the top of her head, a bit to the left then back slightly to the right just before it curled under, next to the neck of her blue Scandinavian design ski sweater. I don't think I knew anything about kissing yet but I clearly remember longing to hold her to me and to feel her cheek against mine.

I was struggling with some new concept in arithmetic in class and the teacher rolled her eyes when I confessed my confusion. Karin came to the rescue, "I'll help him," she said with forced patience as if it were the umpteenth time. She turned around in her chair and bent over my paper making notations and explaining each one, as if I could see anything but her slowly moving lips and fluttering eyelashes. My mother says I brought her to the front door one time and introduced her.

I turned six while we lived in Seattle. My mother arranged to have a small party for me one afternoon. I was elated. I could invite Karin. I waited outside on the front steps for Karin after all of the other six or seven guests had arrived. Finally, my mother came out to tell me that Karin's mother had called to say Karin was very shy and would not be coming. I sat through the entire party in a daze. I opened the gifts and silently cast them aside, commenting only on one occasion that a particular gift duplicated something I already had. I don't remember any of the other kids who were there but I'm sure each and every one of them had a better time than I did.

In fourth grade I fell for two Annettes. I was in the biggest of the post war baby boom classes, the largest class any school had ever dealt with. Throughout my school years adjustments and accommodations were made for us. In fourth grade at Blessed Sacrament the overflow was diverted to an additional class that was split between third and fourth graders. The nun would teach half of the class for a time then give them a busywork assignment to complete while she attended to the other half. The classroom was split right down the middle, third graders on one side, fourth on the other. I sat on the dividing line, next to and just ahead of third grader Annette Kowalski. Annette had pink framed cat eye glasses which made her stand out from the other girls who all wore the standard navy blue jumper uniform with light blue puff sleeved blouses. When Annette would drop a ruler I would fly out of my seat to retrieve it and would embarrass her terribly by planting a fervent kiss on it before handing it back to her. As my affections were far out of proportion with what was normally expected from fourth graders, they were not only rejected by Annette but ridiculed by Sister Assumpta. "Stop smiling. Your face is going to wrinkle up like a raisin." My needy feelings and behavior also invited more of the taunting and beatings and further fueled my sense that I was different from other children, something I was increasingly learning to keep inside.

As Annette Kowalski had failed to accept my awkward advances, I soon redirected my feelings to the other Annette. After making my way home from school by way of the gauntlet at the bus stop, I would often play by myself in the woods at the edge of the neighborhood. I could easily imagine what it was like to live in the woods like the local Lenai Lenape Indians had. I practiced moving quickly and quietly along the trails learning to walk on the outsides of my feet and watching the trail carefully for twigs and dried leaves that might make noise. I learned to sense the animals and birds alerting to other kids entering the woods and could easily conceal myself from the bullies even when they were very close by. When it got dark I headed for our house at the edge of the woods, sometimes having to sneak home to avoid another confrontation with the neighborhood bad guys.

Many nights I hid out until dinner time in the partially finished basement that was home to our recently acquired first television set. The other Annette was there on the tube most afternoons, one of the regulars on the Mickey Mouse Club. I can still see the exact shape of her lips and her pretty dark eyes.

Throughout the grammar school years, I fantasized constantly about being with one girl or another. While Annette Funicello was not someone I would likely end up with in reality, she was not much less attainable for me than the other Annette or many of the others I thought about sometimes day and night. Betty Milford was three years older than me and was the patrol girl on the bus. Annette Kowalski's older sister Mary was another knockout that was out of my league and even as the girls my age began to develop some curiosity about boys, I was the last one they would look at. I was the shortest, skinniest and one of the youngest kids in the class all the way into high school.

By the time I reached eighth grade the consistent rejection of my affections had caused me to hide my feelings for girls deep inside along with the fear and anger that had to be there from all the mistreatment I had suffered.

CHAPTER 27
Sex

My sex drive was another area that I was walled off from. When it started to resurface during my recovery, the feelings were far stronger than I expected them to be. My first experience of sexual climax had been at age eleven. It was not with girls, not with boys and not even from masturbation. It came as a complete surprise and I had no idea what had happened.

When I was eleven I was expelled from Catholic School at the end of the sixth grade. I began seventh grade at a public school, Fisher Junior High. It was a fresh start. All the other incoming seventh graders had come from several public grammar schools in the township so we were all on a pretty even footing. It was the first time we experienced changing classrooms and teachers throughout the day and the sense of freedom from that alone was wonderful. I wasn't being bullied and beaten up by other students or being brow beaten by the nuns. I found I was one of the brighter students and quickly developed several friendships with both girls and boys in many of my classes.

I was still one of the very shortest and certainly the skinniest kid in the school but even in gym class I was able to stay with the pack. We played flag football, wrestled, climbed ropes and learned to use gym equipment including flying rings, vaulting horses, parallel and horizontal bars. The lightness of being I felt being away from the nuns and the bullies energized me and gave me enthusiasm. I had also been through enough already that even though I was among the smallest, I was among the last to show fear in activities like wrestling, ju-jitsu and even football.

Our gym class was comprised primarily of seventh graders but there were several much older kids from the Special Class. Some of them were far larger and several years older than any of us. Playing flag football outside on the athletic field one day all the other kids on my team refused to center the ball against one of the larger kids from the special class. I immediately stepped up and offered to take on the job. On the very next play I was promptly run over and stepped on. My thumb was broken and disfigured in a way that caused the girls, boys and even the gym teacher to wince and turn away when they saw it.

I ended up in the hospital where I had to undergo surgery under anesthesia to put things back in place. I was told that my thumb might not grow in the same way or at the same rate as my other one. The mishap failed to dampen my enthusiasm for school or for gym and I continued to try my best. When it came to rope climbing, I immediately grasped what was being taught about using the strength of my legs to push myself up the rope rather than trying to use arm strength to pull up. I excelled in this area and could easily beat the bigger kids, including the bruiser from the special class to the top of the rope.

Back at home in the woods, the neighbor kids, including the bullies from the Catholic school, had set up what we called a monkey swing which was a rope hung from a tree limb a good fifty feet off the ground. We would take turns climbing a smaller tree at the edge of a clearing, getting hold of the monkey rope and swinging out over the woods. Some of the bullies began to try and show off their strength by climbing the rope. They had limited success as they did not have the proper technique and were trying to climb using mostly their arms. When I announced that I could do what they could not I was met with disbelief and with the expected dares and challenges which I quickly accepted.

I quickly and effortlessly ascended to a point five or six feet below the limb where the rope was fastened when quite suddenly a tingling sensation in my loins grew and grew and grew...into my first sexual orgasm.

There I was almost fifty feet above the floor of the woods, hanging on for dear life and panting and grunting. There were a good twenty kids below watching my climb who now, seeing me frozen in place, gasping for breath assumed I was exhausted. Many of them were afraid I would fall and they began shouting encouragement. "Go Tommy, go!" "You can do it!" "Keep going!"

Eventually the feeling faded and I easily continued the rest of the way to the top then descended back down into the woods and into complete confusion as to what had just happened.

I was physically immature for my age and although I had a sexual climax I did not ejaculate. I had no idea what had happened but being of an inquisitive mind and having most certainly enjoyed the feeling I went about trying to understand and duplicate the experience. I decided it must have had something to do with my legs slipping against the rope so I found a twelve-foot length of gas pipe which I thought would offer less friction than a rope and would make it easier to achieve the slipping motion I suspected was necessary. I stood the pipe upright like a fireman's pole and lashed it to a low hanging tree limb. I climbed the pole again and again. It took several attempts but perseverance finally paid off and I was able to achieve another orgasm. I still had no ejaculation, nor would I for three more years, and I had absolutely no idea what was happening. I didn't know this had anything to do with girls or even anything to do with my genitals. The orgasms I first had were intense and although I knew they started with a tingling between my legs they were so complete that I didn't even know where they were centered.

A few weeks later I was in a tree fort in the woods with one of the older neighborhood boys when he started to masturbate in front of me. I was shocked and said, "What are you doing?" He replied quite casually, "Gettin' the feelin', here let me show ya,'" and he quickly unbuckled my pants and began to show me what the great mystery was all about.

It wasn't long after this incident that a man who was the lifeguard at the Trenton YMCA offered to pick me up on Saturday mornings and take me to the Y where I could swim for free. He bought me cupcakes and sodas and made me feel like a VIP at the Y. At a certain point on the second or so visit he insisted that I had to take off my bathing suit when swimming. This wasn't completely without precedent at the YMCA, where in the past boys swam naked or used tank suits provided at the facility for free. Many boys who used the Y were not wealthy and this was a way to put everyone on an equal footing whether rich or poor. It was originally a Christian organization and this was seen as the Christian thing to do.

The problem was that while the lifeguard was able to explain this supposed requirement in a way that made sense… it did not feel right! And it seemed especially strange because I was the only one swimming in the pool. I got the same creepy feeling that I'd had when walking to school in Castro Valley, CA at age four when I thought that man in the car was watching me surreptitiously. The lifeguard was named Roger. He was a student at Trenton State Teachers College and he was in the Physical Education Department. He drove a little Nash Metropolitan. I could feel him watching me from the glass window in the lifeguard station where he sat hunched over staring at me. I had the same dirty violated feeling that had made me walk the long way to school in kindergarten in California.

I hadn't connected up the dots about the masturbation incident yet, and I didn't have a concrete idea about what was going on, only that it felt very, very creepy so I stopped going to the Y with Roger. Not too long after that I began to hear kids at school start to talk about homosexuals and the like and became quite guilt ridden over the masturbation and even about having gone swimming with the lifeguard.

Once again there was no one I felt I could talk to so I merely suffered through the pangs of cringing guilt, eventually denying my own feelings and pushing them deeper down inside of me and away from my conscious mind.

It wasn't too much later that I began to have fantasies about performing oral sex on women. I had no idea where these ideas came from and in truth I didn't have but a vague idea of what female private parts really looked like. My thoughts in this area were pervasive, intense and often distracted me at school from whatever studies I was supposed to be focused on. At that point I had no idea whatsoever that this was something that people actually did so I kept these things entirely to myself assuming no one else could possibly have perverse thoughts like mine.

Communication with my parents was non existent except for being told what not to do many times daily. I had little trust in my peers having been beaten and bullied so often. The one boy in the neighborhood who I had become friendly with was the one who had introduced me to the masturbation and my guilt over this had caused me to end that friendship.

Another factor in my lack of close friendships may have been the frequent moves and changes in school systems that I experienced. The one good friendship I had begun to form with Eddie Kramer in Castro Valley was hindered by the school officials separating us into different Kindergarten classes.

Later in therapy as I worked at healing from the schizophrenia, I learned that my sexual fantasies and my guilt feelings were not at all unusual. I was taught that a majority of boys engaged in mutual masturbation and even other homosexual activities before maturing fully and moving on to heterosexual sex, as most did.

I went through my teen years and into early adulthood with virtually no understanding of sex as anything but something secretive and dirty. My Catholic school education, the emotionally repressive atmosphere in my home and the pre-seventies sexual mores probably all contributed to those ideas. An Army buddy of mine once observed that the only time I seemed to want to have sex, which at that point usually meant prostitutes, was when I was good and drunk. I was twenty years old and on R&R from Vietnam before I first caught on to the idea that women had orgasms as well.

At twenty-five I was still painfully naive about sex and my therapist was really the only person that I trusted enough to talk to about the subject. Dr. Jane told me she had been sexually active from a young age and I think she found it "charming" if not downright humorous that I was so uneducated about the matter. She was adept at making me aware of choices I had not thought of without making it seem like she was telling me what I should do. One day as an aside she casually mentioned two books that were in vogue and as a result I bought The Sensuous Man and the later The Sensuous Woman. I was shaking when I took the first book to the checkout to pay. The cashier treated it as if it was any other book and not only that but made no attempt to draw me out about why I was buying it as my anxiety had made me fear. Reading both books was yet another level of relief. I found that people not only had thoughts like mine but were freely enjoying all the things I had thought of and much, much more! As I began to accept my thoughts about sex, my feelings of arousal came back with an intensity that was surprising to say the least. This was another indication of just how dead I had been to what was going on inside during the disease and possibly before.

The process was not unlike what I was experiencing as I became aware of the other types of feelings that I had long repressed. I would sometimes get a hint as to what was coming through a dream fragment from the night before. It might seize my attention at some point during the day. I would find myself immersed in some unfamiliar and often disturbing thought or feeling while I was painting or drawing in the semi hypnotic state those activities brought on. I would bring those partially conscious sensations to my therapy sessions with Dr. Jane. She would unfailingly offer support and safe space in which to rediscover and accept the parts of me from which I had long separated or never known. In addition to her sense of support and confidence in my ability to heal, she had a highly developed sense about where to focus to help me break through to the most important material.

As I worked through many terrifying and bewildering discoveries in therapy I began to develop more of a sense of confidence that I could handle these things on my own. This is when the next level of healing began.

CHAPTER 28
War

Dr. Jack Ward had written to the officials at the Veterans Administration that he believed my feelings of unreality began in the war zone and that I had started drinking heavily to suppress those symptoms. Those officials had assigned me a ten percent service connected disability rating. Dr. Ward wrote a second letter stating he felt that was quite low considering my inability to support myself and my tendency to slip back into active symptoms. The Veterans Administration in New Jersey was later adjudicated to have deliberately underrated disabilities in that era to save money. The Vietnam War was so unpopular and veterans of that war so unlikely to be vocal that it was easy to ignore our needs in favor of the veterans of the "Great War." No one acknowledged that two thirds of Vietnam vets had volunteered while two thirds of those in WWII were draftees or that there were twice as many vets who saw action in Vietnam as in World War II. Of those who were in infantry units, the Vietnam vets averaged six times as many days in combat in one year as WWII vets in the South Pacific saw in four years!

I was among those Vietnam veterans who were low priority. As I suffered through my schizophrenia I got no help or support directly from the V.A. beyond twenty-eight dollars a month disability compensation and prescriptions at a reduced rate. I had no contact with other veterans or with doctors who had war experience or even any military background, as far as I know. Dr. Ward was not particularly interested in the origins of my disease. He believed that altering my brain and body chemistry was the key to bringing my mind back into balance and he spent no real time on any focus beyond that.

When material from the war finally began to surface, initially through my nightmares, I was staggered by the intensity of the feelings and by the stunning clarity of the images that were triggered by those memories. Even as these memories emerged I had not yet a real understanding of how much the war had affected me.

I was still nineteen in September of 1967 when I arrived at my unit in Vietnam along with four other young Military Policemen that I met in Fort Lewis, Washington on the way over. By five o'clock that same afternoon one of them was on the way home paralyzed for life. Then just after dark there was a blast a short distance down the street. Several members of my new company were wounded along with other Americans and Vietnamese civilians in a grenade attack on a restaurant. That was day one. I would spend three hundred ninety-four more days in Vietnam and would turn both twenty and twenty-one before I returned to the United States.

The first real indication that I was carrying deeply hidden material from the war came when I began to remember the nightmares about the firefight in which my rifle jammed. I had no idea why I cried so uncontrollably when I first remembered the dream during a therapy session. As usual I began looking for a simple explanation or 'reason' to explain my reaction. In therapy I worked to fully recall the nightmare and became clear that it was associated with a real life incident. I then worked to uncover repressed feelings about that. We came up with reason after reason to explain the strength of my feelings about these things. Still, the stunning depth and power of the feelings persisted.

In one view, the core of my schizophrenia was distrust of my own perceptions. I was not sure whether the world I saw was real or whether it was an illusion or a dream of some kind. The fearful idea that I had already died and was trapped in some limbo or even hell was constantly at the edge of my thoughts. I wondered if at some moment as I was about to be killed during the war I had escaped into some alternate parallel reality. Maybe I had even died consciously but was in some place beyond and in denial, refusing to remember my painful and frightening death.

Out of my close to four hundred days in Vietnam I could remember well over a dozen when the odds were so stacked against survival that it was not difficult to imagine I had died somewhere along the line. The firefight associated with the dream was certainly one of those. There was a Viet Cong unit operating in our area that had been the cause of President Johnson ordering the 1st Air Cavalry into Bong Son. Nixon referred to the incident in his memoirs. A band of Communist guerillas had entered a village at nightfall, gathered the villagers together and forced them to watch as they hacked off the arms and legs of the village chief's children and then beheaded them. Next they gang raped the horrified mother of the murdered children before disemboweling her. Finally, threatening to behead anyone who looked away, they castrated the village chief then killed him. The heads of the entire family were mounted on bamboo poles in the middle of the village and the villagers were threatened with death if anyone tried to take them down.

When the 1st Air Cavalry came into the area in response to this incident, a Vietnamese school teacher ran from hut to hut telling the villagers that the Americans were coming to help them and asking the villagers not to resist. The Viet Cong returned that night and crucified the teacher to a tree, disemboweling her and leaving her to suffer and die.

It was this same band of Viet Cong, thirty strong that later attacked a village, throwing grenades and firing assault weapons into it, wounding more than a dozen unarmed civilians and killing a six-month old baby in the first volleys. My MP partner Darryl Shumaker and I were the only ones in a position to help and we engaged the attackers drawing fire onto ourselves. There were thirty attackers and two of us. It is perhaps still hard for me to understand how we survived. We killed at least eight of the enemy and wounded an additional unknown number according to the military intelligence unit that investigated afterwards. Shumaker and I had to have reached nearly unimaginable levels of performance that night for things to have come out the way they did and it should be easy to understand how I might later question whether we had actually prevailed or whether we were killed or had escaped into some alternate reality or twilight zone.

There were several other incidents that could and did raise similar doubts. I had been thrown out the open door of my helicopter at an altitude of 3000 feet and barely survived by clinging to the gun mount. I had been hospitalized for two weeks on each of three occasions for malaria symptoms and gastroenteritis. On the second occurrence I was bleeding internally and had a fever of 106 degrees. The medic at the aid station said a fever that high is often fatal and he told the other troopers being medevaced with me that I could die on the way to the hospital in Quinhon. I lost consciousness moments after hearing that. I had not an ounce of strength left to fight with and I remember thinking there was nothing left to do but let go.

In another incident, I was caught in a powerful current while recovering the stiff and rotting body of a fellow soldier from the river at Bong Son. I became exhausted and can remember feeling quite detached from my body as I tried to try to swim sideways out of the current while gagging from the stench of the corpse.

I was caught in two major ammunition depot fires and I remember in each case thinking I would probably not survive. I lived in a small sandbagged outpost surrounded by a dry moat that served as a Vietnamese police station. It was attacked every single night by everything from snipers to sappers. Again I can remember time after time when I felt death was likely.

With my perceptions in question and my less than solid grasp on reality, emerging memories of these and other similar instances strengthened my fears that I was dead or hanging somewhere between life and death and only imagining I was still alive. The amount of killing I had seen and the terrorism perpetrated by the communists against their own people also kept leading me back to the hard to ignore conclusion that evil was the ultimate ordering principle in the world in which I was trapped.

I assumed that repressed fear would be the dominant discovery as I broke though to my war memories. In spite of having lived through many situations that could have taken my life, I felt precious little conscious fear when I was in Vietnam. That may have been due to the way I learned to sublimate fear during the years of beatings I suffered as a child, but I had another theory as well, which I had discussed with some of my buddies over there. I said I could always tell new guys that had only been in country a week or two because their faces were red and they talked too much out of nervousness. After a few weeks they slowed down, the redness disappeared and their movements became more focused or directed. They talked less and listened and watched more. The nervousness had been controlled. I speculated at the time that when we first arrived in country our senses were on overload and we felt fear as we should. After a couple of weeks, we intellectually understood what could happen to us but we no longer felt the full effect of our fear. We could not afford to and still function in our duties.

I assumed that my fear of dying, of being paralyzed or losing limbs, or of being castrated, which was commonly done by the communists to captured and wounded Americans, had to have been repressed and would be the primary feeling that would resurface as I faced the memories of Vietnam.

Surprisingly, most of what came up was guilt or shame about not having performed perfectly in situations. I would turn things over and over in my mind wishing I had been quicker to respond in a certain way or quite oppositely that I had been more thoughtful and circumspect in other situations. I did process enormous, nearly unimaginable amounts of fear during my recovery but much of that I associated with my anxiety over evil and my fate in eternity. When I became open to experiencing fear over combat and other situations from the war, less fear surfaced than I had anticipated. Instead I spent days, weeks and months second guessing how I had performed in difficult and sometimes impossible situations. Massive amounts of guilt and regret were what came up.

There was a particular event that haunted me on a moment to moment basis for a full three months. There was an attack on a bunker manned by my fellow MP's one night and several of us rushed to the perimeter of the forward base camp to assist. There was a fatality in the next bunker. Friends of the soldier who was killed thought it was due to friendly fire, our fire; that the boy had been hit after the Cong had stopped attacking. The report by the medics who inspected and registered the body exonerated us saying that the puncture wound in the soldier's skull was the wrong size to have come from one of our weapons.

That rational explanation did nothing to quiet my feelings of guilt. I spent countless minutes, hours and weeks playing back the incident in my head. For three full months I could think of nothing else, or talk of anything else in the therapy sessions. I tried to remember where each and every round fired from my rifle was directed. I could picture the red streaking paths of my tracer rounds as they started on the lines I intended but in my anxiety and guilt I could never get a clear picture where some of them impacted. The obsession went on and on. Moments after I awakened in the morning the anxiety would surface and my mind would kick into runaway mode trying desperately to picture the whole situation clearly enough that I could free myself from the awful possibility that I had killed some mother's son that was one of our own.

The torment continued on and on. Finally, late one night after lying awake for hours unable to block the thoughts, I climbed out a window and sat high up on the roof. I looked out into the darkness without distractions and began once again to reconstruct the occurrence moment by moment as best as I was able. I remembered coming up over the berm behind the bunker and starting to fire, I remembered one of the MP's raking an area just beyond the perimeter with machine gun fire. I could see my rounds going out beyond where the attack had come as I tried to bring fire on the trail I thought the VC would take as they pulled back. I could see muzzle flashes from the bunker to our right and I remembered several of us pointing at what looked like movement just beyond the no man's land along the perimeter, and firing into that area. It was at that point the images became unclear as to where my rounds went and it was moments after that a voice from the bunker next to us screamed, "Medic! Medic! Medic!" over and over, hopelessly against the long dark night.

Never able to fully remember the last moments leading up to the mortal wounding of the boy in the next bunker, that night on the roof I finally gave up any hope of clearly convincing myself that I was not at fault. Accepting the feelings of self loathing and revulsion that went along with my possible guilt, I then tried to visualize the thing I had most resisted, that my own rifle shots had hit the opening in the next bunker. I thought, "If I killed him, so be it!" I had struggled so long against those thoughts and now as I let go into them, a strange thing happened. Not only did I fail to visualize my rounds going where I didn't want them to, as I let go of the resistance, a clear memory of the gap returned and I became quite sure that it was not me who had caused the death. It had taken three months to learn all over again that the things I resisted most were the very things I had to accept and face fully in order to be free of them.

The firefight against the VC platoon who attacked the unarmed village brought up a different issue. I had experienced fear in the nightmare when my rifle jammed and extreme fear at the moment in the dream when the attackers ran through my body as if it wasn't solid. The most powerful and reoccurring feeling associated with that night however was grief. It wasn't sadness for the civilians who had been wounded by the enemy nor even for the one civilian in a nearby hut who I believed we had accidentally wounded as we fired at the enemy. No, again and again I was overtaken by my own tears and deep sadness when I pictured a group of enemy soldiers that I had personally and surely killed as they crossed the dirt path toward the village. Logically I could convince myself that I should feel no regret. These were the people who had already committed horrible atrocities and again were in the process of killing unarmed villagers and their children. Surely the Vietnamese and American authorities had praised us for how we handled ourselves. There was no rational reason for me to feel any regret, yet I could not think of the look on one Viet Cong's face as I shot him and his two comrades nor about the piles of bodies of dead and wounded lying in pools of blood in the thick dust in front of our bunker without crying profusely and being overwhelmed by deep and profound grief.

After many attempts to explain away or justify those feelings without success I came to the conclusion that this was one set of feelings I was going to have to live with from time to time. The good news was that I felt more able to do so and this was another giant leap forward in my growth.

Years later I visited with Shu, my partner that night, and he told me how he had struggled for years with many difficult feelings also. In his soft spoken Arkansas drawl he shared a piece of wisdom that was finally helpful. He said simply, "When ya' kill somebody, I think it makes ya' sick."

My war experiences raised several challenges, some of which I would continue to deal with for years to come. Many of those problems were shared by other veterans who did not become psychotic, but had difficulty adjusting or coping with memories. As far as my own schizophrenia, my work at accessing memories and feelings from the war shed light on two significant areas.

I spent a lot of time second guessing how I had reacted in many critical situations. I often came away feeling as if I had done something wrong or could have done it better. For much of my earlier life I had gotten a constant stream of negative messages and criticisms about my behavior to the point where I became an extremely internal or introverted person. I had an over active imagination often fantasizing about doing things great and small but I seldom carried out even the simplest of those plans for fear of making a mistake or getting punished in some way.

In desperate situations in the war I did not have the option to stay inside my head and imagine the right solution. Rather I was forced to act on my wits without time to second guess myself. It was interesting to see how the negative voices I had internalized were with me to such an extent that they caused me to look back and feel guilty even when I had done my best in situations where there was no other choice but the one I took.

The second major revelation in my therapy was about my sense of reality. Dr. Ward believed my feelings of unreality began in the war zone. That certainly made sense and there were scores of incidents that I witnessed or participated in, which certainly would seem unreal to someone who had not been to war. Many of the things I experienced and witnessed did in fact cause a sense of awe in me, but in truth those things, as painful and sometimes frightening as they could be, were some of the most real moments of my life. They were so real and so urgent that they forced me to come out of my head and react with my entire being.

I began to see that my sense of unreality may have come about in a very different way. When I returned home to "the world" as we referred to it,

I came home to a place filled with people who drifted along in complete ignorance of the reality I had just experienced. Worse, many seemed selfish and full of deception and greed. They were overly concerned with the most trivial problems and seemed to have no regard for the suffering that was still going on daily in Vietnam or for that matter anywhere outside their own myopic fields of view. They had no ability or motivation to work together to solve problems and even less perspective on what a real problem was.

I began to see that it was this world of so called normal people that had begun to seem unreal to me. This world in which people lived in cocoons of ignorance and self absorption was the one that I had wanted to escape.

CHAPTER 29
I Can See Clearly Now

I did not go to the movies during the worst of my disease. Just prior to my psychotic break I had seen the Stanley Kubrick movie 2001 A Space Odyssey. It had further fired my mental gyrations and fears about the nature of reality and the origins of the material world and of consciousness. Now as I found myself more able to handle my fears and other feelings, I began to slowly reengage in activities that had for some time been overwhelming. For awhile I had shopped at the Pathmark twenty-four-hour supermarket after midnight to avoid crowds. Now I started to shop more during daylight hours. I was able to hold down jobs for at least for some length of time and I became more involved again in racing our 1957 Chevrolet "Cherokee" with my partner. I was also working part time at the drag strip.

Finally, I began to date tentatively and I started to venture out to restaurants and even to a movie. The first movie I saw was the Last Picture Show. It was a Hollywood adaptation of a Larry McMurtry novel about a boy who comes of age in a dying town where most people live lives of fantasy, self deception or quiet desperation.

The main character in the movie becomes disillusioned and hopeless. He gets into his pickup truck and drives wildly out of town looking for escape from his dismal and disappointing life in the small town. He finally pulls to the side of the road and after reflection turns the vehicle back toward town. He has nowhere else to go.

In my state of mind when I saw the movie, I interpreted that scene in quite a different way. I was convinced that when he sped out of town in his truck he intended to kill himself by crashing into someone or something else. I believed when he turned back toward town it was because he had lost the nerve to end his depressing life.

When I left the movie that afternoon the weather was gray and rainy. I was numb as I drove to my apartment. I climbed the stairs and flung myself face down on my bed. My chest began to heave and I began to sob, to wail and to shriek. I cried for literally hours and on into the night. I cried myself into exhaustion and would lay motionless for a time until the next wave of hopelessness would come over me and I would begin again to cry uncontrollably. I stayed in that state for a full day and for most of the second day I continued to cry on and off. I never left the bed except to use the bathroom. The feelings of darkness, of despair and of what I can only describe as primordial sadness were so strong that they blocked out any other thoughts. I was so lost in the feelings that I could not imagine going to work, eating a meal or even ever leaving the bed I was lying on. I did not think of calling my doctor or of anything beyond the very powerful feelings of the moment in which I was lost.

As I had learned in my encounters with the deeper, seemingly primal levels of fear and anger, the answers to my pain did not lie in escaping from those feelings or in explaining them away. The answer in this case was the same as before which was to accept the feelings and to stay with them. The effectiveness in this approach is not easily explained. Initially when uncomfortable feelings surfaced during my recovery, explaining them or justifying them was important. Having a mental or cognitive "handle" or explanation for the feelings allowed me to stay in them longer. As I did I learned that the feelings would not kill me or as I had feared, cause me to dissolve, leaving me exposed and naked to eternity or to a kind of ultimate judgment from which I feared further torment or punishment.

As I became able to penetrate deeper and more disturbing feelings, what I found was not that I dissolved but that my ego, or the thoughts and judgments that I had internalized over the years, that is what dissolved. When that did happen, instead of feeling like I had lost a part of me, I felt like I had grown larger. I felt I had awakened to another aspect of myself that had been there all along. Each time I let go of some resistance or some idea of the way I thought I should be or the way things must be, I felt freer, more solid and better about myself and about the world.

This is likely the most important discovery I made in my schizophrenic process and remains to this day the most important discovery of my life. The ability to penetrate my feelings, thoughts, and habitual patterns was learned slowly. Initially I was able to experience uncomfortable feelings, images and thoughts in somewhat semi conscious states as much of my mind was elsewhere, for example following the storyline of an old movie as I sat painting. Access to many of the deeper disturbances first came up for me while I was dreaming.

I used those access points or impressions to explore the deeper, fuller feelings in therapy with Dr. Rittmayer. In the beginning I was merely looking to explain or justify the feelings. As I stayed with the feelings, with the intent of trying to understand them, they emerged more and more powerfully to the point where they overwhelmed my ego, or the thinking part of me that was trying to get on top of them. The first time this had happened was when I began to remember the dream of the firefight when my rifle jammed.

It was in this phase that I truly experienced the magic of healing. When I fully gave myself to the feelings, they were so powerful that they caused me to experience myself beyond thought, beyond ego's interpretation or explanation of the experience. My experience became excruciatingly direct and in those moments the feelings that had been rejected, repressed and denied were transmuted into something quite different.

I began to see them as basic energies that could be used in a positive way. Fear or more accurately infinite paranoia or terror, as I had experienced it, became something quite beyond fear. By moving into and beyond the fear I found an ability to stay in situations that I would have normally run from and to see those situations so clearly that I could function in them accurately.

When I experienced my anger more fully and directly, rather than reacting violently or denying or justifying the anger, I found a sense of stability or immovability in me. I found, for example, that I didn't need to be drawn into other people's anger or to lash out in situations that seemed unfair to me, something which had been a huge trigger for me previously, probably due to the bullying in my childhood.

When I became more able to fully accept or merge with these very strong feelings another extraordinary thing happened. The opposite polarity of these energies emerged quite clearly and powerfully. As I let my fear have me completely I found myself, quite to my surprise, in a state of fearlessness. Anger, instead of motivating an agitated or aggressive response became a sense of clear headed peace. In that state I found a sense of peace and compassion for the difficulties other people were going through. By seeing into my own anger clearly, I could better understand how someone else had ended up in that state and rather than be hooked by their attempts to lash out, I could touch into my own state of immovability or peace and sometimes even just presenting in that way helped them to let go of their own rage.

To be clear, in the beginning I experienced glimpses of uncomfortable feelings in only a partly conscious way. My success in integrating these feelings caused me to make greater efforts to experience them more fully. I did this initially with the support of my therapist. At that stage I was still looking to explain and justify what I was feeling. I then began to develop the ability to sustain contact with my feelings to the point where it empowered me to experience even the most extreme feelings with less and less mediation or management from my thinking mind.

Eventually I became able to make a very great leap of faith and to experience things more directly. The feelings became so overwhelming that my ego or thinking mind was completely incapable of experiencing the extreme primal nature or texture of these energies and it let go, or was wiped out by the energies. On each occasion, when I was able to sustain and experience the feelings, I emerged feeling not less but more of a person. Instead of feeling like I had lost some part of myself, I felt as if I had become something more.

What I had been desperately hanging onto, the thing I was so terrified would dissolve, was not me, but rather my ego or an internalized false idea of who I was, an idea that had kept me from being myself, my true self which had been there all along.

When the feelings of sadness and despair overtook me as I viewed the Last Picture Show, I had learned through my experiences with fear and anger to allow myself to accept the feelings rather than explain or deny them. As I lay sobbing and wailing later that day I had already given myself over to the feelings to the point where I was beyond managing them with my thoughts. For two days I experienced despair and hopelessness at a profound level.

Thoughts like "How could the world have treated me like this?" went through my mind. I wondered how anyone could have allowed me to be beaten up for so many years. I tried to fathom how anyone could perpetrate the horrors of war or injury to other people that I had seen overseas. As I had learned, I tried not to hold onto any of those thoughts or to attempt to resolve any of the questions raised by them. Instead, I focused my attention on the discomfort, the overwhelming sadness. The feelings again grew so strong that they virtually wiped out my thoughts. Finally and fully I experienced the energy directly, so directly that words were and are irrelevant. It was as if the sadness, the aloneness and the despair I felt were too strong too be attributed to the life experiences of one little person. Later I wondered if theories about collective unconscious or race memories could have been attempts to explain similar experiences. I still wonder if some of the more profound and timeless works of art and music were attempts to validate these types of experiences by other people who touched into feelings so deep they defy explanation in logical terms. The best description I have been able to come up with is to say that I experienced primordial emotions.

Sometime during the second day the tide shifted and as I referred to earlier, the opposite polarity of what I had been experiencing began to emerge. I became aware that the sun had broken through the clouds and was streaming through an opening in the curtains directly onto my face. I suddenly felt hungry and I wanted to be outside so I showered and went down to my car. I felt refreshed and alive and very glad to be so. I opened the window in the car and turned on the radio. A song, "I Can See Clearly Now," sung by Johnny Nash was just beginning. It was a song I had never heard before but I will remember it to my dying day.

The tune was still with me later that night as I took out a sheet of artist paper and some water color markers. I sang to myself, "I think I can make it now, the rain is gone," as I sketched a simple abstract figure of a boy stretching his body and arms in exuberance as he greeted the morning sunrise. It was a picture of simple joy.

I took the picture to therapy and I held it out to Dr. Jane. "Oh, Tom," she exclaimed. Without another word spoken between us, we both knew with certainty that the worst of the long dark night of my terrible disease was now behind me.

CHAPTER 30
Intuition

For years I had been concerned about my death, about what would happen to "me" as my body dissolved back into the ashes and dust. Now as I began to discover who the real me was and to feel more solid and stable, a remarkable thing happened. The world around me became more solid and workable! For a very long time the world had been gray and foreboding. There had been fearsome demons lurking in every shadow and in fact the entire universe seemed something of a shadow itself, a veil hiding the great evil that was behind it all. Now a different world was emerging. It was an inviting world filled with light and even joy, a feeling that had been long gone from my life.

I felt as if I had just been born while only weeks before I felt like I was dying and the only hope of any peace was in trying to convince myself that my mind could die with me. I was more and more willing to face life on its own terms and the clarity that came with my breakthroughs helped me feel able to meet whatever challenges were ahead. I realized I had been more afraid of my own feelings, afraid of being my own self than I was of whatever situations I might encounter.

Compared to what I went through in war, in childhood and throughout my schizophrenia, most problems I thought I might encounter now seemed manageable. I came to understand that most of the things I feared were not actual situations but rather projections or fantasies of what might happen or past memories, often distorted, of situations that I had not dealt with well. I began to learn that staying in the present was the key to dealing effectively in the world.

Early in the therapy Dr. Rittmayer had given me a paper she had written. She wanted me to understand the theoretical basis for her approach to therapy. The approach was based in Jungian theory and in particular on a way of classifying personalities that Carl Jung had introduced. Dr. Jane said the therapy would be based to an extent on understanding where a person was out of balance by being too focused in one aspect of their personality. At that time, I understood the concepts but not the relevance to my struggles. What might have seemed more relevant to me at the time were Carl Jung's writings about the collective unconscious as I thought my problems were more about the nature of mind and of reality.

The typology that Dr. Rittmayer referred to divided people into four main groups and each of these groups had a style of relating to situations that was in some way related to time. The Feeling person, for example, was to an extent trapped in the past and tended to relate with situations based on feelings about previous events. That much I could understand early on and I understood that I had tendencies in that area.

The second type was the Thinking person who was often focused on an imagined or projected image of the future or the way he thought things would or should be. Again I could understand the concept and could see how one could make bad decisions based on a wished for or even feared future that might have no basis in reality.

The third type was the Sensation person who could only relate with his or her own immediate needs, particularly physical needs and desires. In the extreme this person had little regard for the needs of others and no ability to learn from his past or to project future consequences. I identified least with this type as I understood even in the early stages of my illness that I was overly cautious, and more likely to ignore my own needs based on fear of future consequences, or past failures and especially for fear of upsetting or angering someone else.

The last type presented was the one I least understood, the Intuitive. Dr. Rittmayer offered a colleague of hers, Dr. Carlton Fredericks, as an example of this type and she said these people just "know" what to do, but other than understanding this as something desirable I did not know what it meant.

What I did learn from her paper and from my discussions about it with Dr. Rittmayer was that her approach would attempt to free me from attachment to feelings from my past that I might sometimes refer to in new situations inappropriately. While that made some sense intellectually, it initially seemed far removed from my central issue which was whether I was in hell already or on the way there.

As I emerged from my deep despair after seeing the Last Picture Show I had crossed one of the final barriers to my recovery. I now accepted the ultimate hopelessness and futility that I so resisted and by doing so I was free to experience the other side of life. If life was futile, if the devil would ultimately take my soul... I still had today!

All the convoluted and knotted thinking I had done to try and figure out the world, all the worrying and anxiety I put myself through had absolutely no effect on reality except to cut me off from experiencing it properly. I began to realize that I had no control over who I was born as or whether I was created by a God or by some random permutation of the Big Bang. Likewise, I had no control over the nature of the world whether it was random, or controlled by an intelligence that might be ultimately good, evil or some evolving balance of the two. All I had to work with was my own self and whatever situation I was in at that moment.

I began to glimpse what it meant to be Intuitive. As I became more accepting of what was, both internally and externally, my mind fell silent, or at least more so. I was no longer desperately checking on past experiences or feelings to try to gauge how to act. I was no longer projecting dozens of possible outcomes before taking an action. I had developed the beginnings of trust in myself and surprisingly in the situations themselves.

The Jungian classification of the Intuitive type was something I now began to appreciate. By not being attached to my projections or to my ego's interpretations of what was happening or what might happen, I was able to see situations more fully and to my great surprise I seemed to "know" how to react effectively.

I found that pacing myself both physically and mentally was the key. If I was moving too fast or too slowly I could miss an important detail of my surroundings. If I was thinking about the past or future, it became obvious that I could not see what was in front of me at that very moment. Now the Jungian Intuitive type that my good doctor had introduced to me had some meaning. If the mind was free of fear and other limiting factors it was open to seeing whole situations as they are. The more I was able to free myself of expectations, fears, and even concepts the more directly I could perceive the world. To my great surprise I began to find that my world told me exactly how to react... accurately. This was the secret of the Intuitive type that I had failed to comprehend.

I began to looking forward to my days and to experimenting with my new found skills in dealing with whatever came up. If I began to revel in my success or to pat myself on the back, I found that took me out of the present and closed off the openness or more egoless state in which I could function at that intuitive level. When I did move at a proper pace physically, my mind would slow, leaving room or open space to see more accurately what was in front of me. The moving mind or discursive thoughts would partially cease and the still part of my mind would simply "know" what to do.

Previously my mind was full of chatter, trying to interpret, manage and control all that I encountered. Now having faced the worst of what I feared might happen, there was little for that part of my mind to do and day by day it started to slip away. Instead a deeper, fuller, more confident and certainly more at peace part of me took its place. I was truly healthier than I had ever been before. Make no mistake. Schizophrenia had devastated me, but now as a result of having accepted that process fully and working with it instead of against it, I had begun to grow in ways that I would never have believed possible during my disease or even before I had fallen ill. Day by day and sometimes minute by minute my new or real self emerged. Layers and layers of my sick mind found themselves no longer useful and simply fell away.

CHAPTER 31
Layers

For the first two years of my breakdown I felt little besides varying levels of fear. My world was one of grays and blacks. The earth was not solid for me. I was in constant dread that it would decay or dissolve around me at any moment and that I would dissolve with it. I was always fearful that the ground could open up beneath me and I would be swallowed by the deep black hell of eternity.

When I awoke out of the hopeless darkness I had fallen into after seeing the Last Picture Show, the world had changed. The Johnny Nash song "I Can See Clearly Now" had come on the radio at what seemed a magical moment. I felt a release and a euphoria that was beyond anything I could remember. The world suddenly seemed a more solid and workable place. I had a sense of my own solidness. I became less afraid that I or my body would dissolve.

The Hugh Prather book, "Notes to Myself," that Dr. Jane had given me some time ago came to mind. The idea of communicating with my own self and trusting my own process began to make sense. My surroundings now seemed full of concrete feedback rather than ambiguous and frightening omens of impending doom. I felt better equipped to deal with life rather than hide from it. It became clearer that cocooning or hiding from the world had merely been a way of hiding from myself.

As I began to reengage with life, I also became more keenly aware of who I really was. I found to my great surprise that I actually liked most of what I found. The parts of myself I did not like were often things I could do something about. If by being more open to my feelings, I reacted too angrily at someone who upset me, I could apologize and I could also learn enough from the experience to do better the next time. Earlier in my life I would have simply ignored or repressed the anger, however justified it may have been, until the feelings built to a point where they were too strong to risk expressing. Over time I had learned to ignore my feelings and to tolerate situations that literally made me sick.

The biggest surprise was that when I brought my real self to the party, so to speak, things went better. I was beginning to discover a basic order or sanity in the world. The evil universe I feared seemed to fade away and instead I found that the world was actually workable. What had been missing from the formula was my own true nature, which I had hidden deep inside so I, or whoever I thought I was, could fit into some concept of the world that I had internalized. The Catholic nuns and priests had taught me to anticipate a world of eternal punishment unless I could essentially conform to their narrow view of what I should be. My family upbringing had done nothing for me except shut me off from my emotions so that I could fit into the overly controlled, unemotional, predictable world that my alcoholic parents were comfortable within. The years of terrible beatings I suffered through had helped paint a picture of an unfair world. The war had only increased my suspicion that the world was random and unjust at best, or more likely evil, and I had tried to reject it as unreal.

Now I began to approach my daily moment to moment life with less fear about what might happen to me a hundred million years from now. As I stopped trying to control or ignore my feelings, I found myself in a different world entirely.

I thought back through some of my earlier questions about the nature of reality. I had the startling insight that all along I had been trying to come to terms with the world through one concept after another. I had tried to understand or explain the world through religious, spiritual, scientific or mystical viewpoints rather than experiencing it directly. The same was true of my own self! I had never just simply been myself. I was always trying to fit my behavior, my feelings and even my thoughts to an idea of who I thought I should be. The true nature of the world as well as my own true nature had been hidden behind layers of filters, concepts and explanations through which I saw everything.

As my fear subsided it was replaced by stillness or open space in my mind. Within that openness I could see those old thoughts and feelings arise and dissolve without attaching to them or letting them energize the fear. I could see clearly how the layers of ideas about how to behave or react had limited me. In small ways I began to step beyond those limits.

It was a leap of faith to allow myself the freedom to react to things without over thinking them. I knew I had been stuck within my thoughts far too long. I began to experiment with small manageable risks.

I would be driving toward an intersection at which I would have to turn one way or the other. In my earlier state of mind, before the breakdown I would have thoroughly thought through the various permutations of turning left which might mean going to the grocery store before the library which in turn might mean having frozen food in the car longer and so on.

During the breakdown those simple choices became laden with concerns and thoughts about everlasting consequences that were extreme. Now I was attempting to pace myself as I approached choices in a way that allowed me to still my thoughts and instead "feel" the right choice. What I found was that my deeper or "intuitive" self seemed to be in tune with things in a way that often guided me to the right place at the right time in ways that my thinking mind or emotionally reactive self could never have done.

Trusting my gut and going to the grocery store before the library, without submitting to the worry over the frozen food, turned out to be the right thing after all. I ended up running into a friend from grammar school who would not have been at the library a half hour earlier. She and I became friendly, began to date and stayed lifelong friends.

This was one of many small occurrences that caused me to pay more attention to my newly discovered intuitive self and to be less ruled by the obsessively thinking self which I was finding out was no more than a collection of thoughts and feelings that I had hung onto for various reasons. These thoughts and feelings had been layered one upon the other, filling my awareness to the point that they obscured the basic natural awareness that was the real me!

At the local library one day I was struck by the idea that someone could become so lost in looking through the overwhelming number of books there that they could fail to have any sense of the original space that held them. Another analogy I thought of was the hospital. Although the original building was created as a healing space it was now so full of sickness that it was hard to see anything else there.

I continued to discipline myself to stay more centered in my own "original space." As a result of having processed many long repressed and overwhelming feelings through my work in therapy, I became less afraid of losing myself in them. They simply became part of my awareness as did my thoughts. I was able to experience both of these things along with my sensory perceptions as parts of a balanced view of my world. Prior to this time, I had no idea what to trust. My thoughts, feelings, messages or rules learned in my upbringing, and my perceptions of the external world had all seemed at odds. My internal conflict had been so extreme that my external world and my own psyche both became places that were unlivable.

Now I was essentially starting from scratch. My old knotted, convoluted and confused personality had broken down and was essentially ceasing to exist. It was excruciatingly painful work but I had finally abandoned the defense mechanisms that had kept that old sense of self in place. I was becoming able to experience thoughts and feelings that previously frightened me so much I had imagined a fearful world of evil and doom to keep them hidden from me.

A truer, simpler, deeper and more confident self was emerging. What was falling away was only a false sense of self that I had created to make myself into what I believed was acceptable to others. It was instructive to look back at that mechanism once I began to leave it behind. Many of the behaviors, thought patterns, and habits I had incorporated were to appease people or systems that I actually had far less than high regard for. I could see how my parents' unemotional and predictable life worked for them, but I had known all along I needed something different. The view of the world purveyed by the Catholic nuns might make sense to them but from the very beginning it had raised issues for me. I had been physically beaten up by bullies so much that I had learned to hide out or fly under the radar so as to not attract any attention to me.

Some of the self limitations that I discovered were small but each one was significant to me as each one clarified the mechanism that had separated me from myself. Each layer I pierced immediately gave me an expanded sense of self. With each small step I felt as if I could breathe more deeply and that the air itself was healthier and more invigorating.

There was a disc jockey on a radio station in Philadelphia that used to frighten me. He would talk in low spooky tones about things that were probably meant to send a paranoid chill through people who were high on drugs. I had become convinced he was a mouthpiece for the great evil and that his messages were directed at me personally. For a time, I had to stop listening to that station entirely. This is one small example of how I limited my life to the point where I lived in a kind of cocoon. As I broke out of the cocoon and found even the worst of my fears were manageable after all, I began to risk facing them one at a time. I became bolder. I not only turned the radio station back on but I went to an event in which the DJ was playing in a softball game. One look at this little nebbish awkwardly swinging a bat or fumbling and dropping a ball was more than enough to instantly deflate the idea of the supernatural devil's agent I had projected onto him.

Not all of the layers that had limited me had to do with the schizophrenia. Some were simple. After years of being bullied and kept from playing sports and games by the larger kids I had come to believe I was incapable of competing in those things. For no other reason than to test that limit I began to play softball and found to everyone's surprise that I could hit the ball as far as anyone on the team due to my physical flexibility and to my mental ability to grasp technique.

I took a Red Cross lifesaving course. It was another of those intuitive decisions that went beyond any logical reason. I saw an ad and just decided to do it. I was still thin and physically weak from my illnesses in Vietnam. I overheard the two women who taught the course talking about whether I would be strong enough to complete the requirements, but I kept working at it. One day I became exhausted while swimming many, many laps of a test. I mentally hit a kind of wall. I thought for a moment I would die if I did not stop. A feeling of separateness from my body came over me and the pool suddenly felt dirty and dark. I thought I would wretch or vomit but my new self refused to listen to the feelings or to my desperate thoughts and I swam on and on.

I was numb when I climbed out of the pool and a few minutes later when we were dismissed I went to my car and began to cry. As the tears flowed I realized I had just experienced a flashback of an incident in Vietnam when I nearly drowned retrieving a dead soldier from a river. I let the feelings and images continue to wash over me. I remembered feeling separate from my body as I fought the current and the weight of the stiff and waterlogged corpse and gagged from the stench of the rotting flesh and the filthy river. The feelings stayed with me for a good part of the day and I dreamt of the incident a few times soon after. When I returned to the lifeguard class I felt less tired as I swam. My strokes slowed down and became more powerful simply by letting go of a layer of tension that had held me back because I didn't want to remember the difficult incident in my past. I almost instantly became a stronger swimmer and later worked part time as a lifeguard to supplement my income.

Throughout that third year since the beginning of my schizophrenia I worked at learning to trust the world and my own reactions to it. I did not achieve success all at once but in small steps one day at a time. There were ups and downs. Overall I felt like I was making progress. Sometimes I felt I had reached a new plateau in my growth. I knew I would continue to face challenges and setbacks but at the new level of growth I felt sure that I would never again fall as far. Dealing with the hopelessness I felt after watching The Last Picture Show brought me to such a plateau. Facing the terror and anger that had surfaced through my nightmares gave me similarly increased levels of confidence and strength.

Now I was approaching small life events more bravely. As I did, limiting thoughts and fears would surface. At this stage I was able to see through them rather than react automatically. In many cases I could see that the restrictions, instructions and burdens I carried with me served no useful purpose and much of that baggage was consciously left behind.

CHAPTER 32
Therapy ends

Dr. Rittmayer began to suggest that the work we had set out to do was nearing completion. She said the work was now something I was capable of handling on my own. When she hinted we should plan to end the therapy sessions I didn't immediately resist. I did ask if instead of ending them I could continue to see her less often. For some time, I had been coming for two hours once a week. I asked if I could possibly come just once a month for a while. I agreed with her that I had become much more capable of dealing with my life but I still felt I needed her support to fall back on. She flatly refused. We decided to put off the decision until a later date and to continue on the full weekly therapy schedule until that time.

Only two weeks later I began to chuckle during a therapy session. The doctor asked me what was so funny. I honestly confessed that her efforts at helping me to focus or at giving me guidance had suddenly become so transparent that they were funny. She took this in stride and said this was proof that I was ready to move on as she had suspected. I couldn't argue with her and no longer felt the need to.

After about two and a half years of working with Dr. Jane Rittmayer regularly and intensely we agreed to end the therapy on the spot without any further ado. She had truly saved my life. Her therapy techniques and her education were certainly helpful and I had developed a high regard for her abilities. What had allowed me to trust her initially was a different matter. She had shared with me her own dark secret that she herself had been schizophrenic. That was possibly the single most important moment of my entire life.

About the same time I stopped seeing the therapist, I experienced one of those events that can be explained as either coincidence or fate depending on one's view of the world. I had started to learn that my "views" of how the world worked often kept me from seeing the truth of situations and so I was trying harder not to judge or resist events that didn't fit such views perfectly. When I took an elementary film course at Trenton State College my instructor briefly mentioned a summer course in film production he had taken at New York University. Right after the therapy ended I was at the Mercer County Library dismantling an art show I had on display there and found a poster advertising what looked like the very course the teacher had mentioned. I was instantly convinced this was the next right move for me.

I sold everything I could sell including my Volkswagen to raise money. I broke my apartment lease and moved the remainder of my belongings to my parents' basement. I left for New York with enough money to pay tuition for the six week course and just barely enough to pay for my room and a minimum of meals at the university dormitory. Off I went with one small suitcase containing an extra pair of jeans, a few t-shirts, underwear, toothbrush, razor and a supply of vitamins and Vistaril.

When I arrived I was told I hadn't actually secured a spot in the class. I had given up my apartment, my car and many of my belongings and I was on a waiting list. I was convinced I was on the right track so I gutted it out and I began to attend the classes without being formally admitted. On the third day the instructor began to break the class into crews of four and realized there was an extra person. When he questioned the class I owned up to the fact that I was the extra person. Not more than an hour later, the phone in the studio rang. It was the admissions office conveying the fact that someone had just dropped out of the class. They told the teacher to tell me to come by with a check as I had been accepted!

The experience at that summer session set the stage for my career. I loved the work and I loved being out among people rather than sitting alone painting. I excelled at many aspects of filmmaking and was so involved in the process and in dealing with the moment to moment realities of the big city that I had no time to indulge my fears of the nether world.

I made several friends including another Vietnam Veteran named Tony. He had been a combat photographer during the war and we ended up working together on several film projects. We went along with a couple of other friends from the class to see a movie at a theater on the Upper East Side. The movie was called Electraglide in Blue. It was about a couple of motorcycle cops in

Arizona and it was kind of a reverse *Easy Rider* in which the hippie drug dealers turn out to be the callous violent ones. There is a point in the movie when one of the policemen becomes unbalanced, gets drunk and starts firing his weapon in a trailer park putting the other inhabitants at risk. To protect the neighbors, the hero, played by Robert Blake, has to kill the man who is his best friend and partner on the police force. The scene was particularly bloody. I began to cry and could not stop myself. I could see the other friends we had come with and the other people in the theater turning to look at me. I realized none of them had any clue as to why I was crying, except Tony. Tony, like me, had seen useless death and bloodshed and he knew all too well that I was not reacting to an image on the big silver screen in front of me but to a memory of another useless death and I was feeling the loss of good people for no good reason. Tony reached over and put his hand on my wrist and said quietly, "It's just a movie, man… just a movie. When Dr. Jane told me she'd had a breakdown, I knew she could relate to what I was going through. Likewise, I knew Tony had some idea of what was happening to me and it helped me to calm down.

The rest of my time in New York went smoothly. I completed the course and took a bus back to Trenton with only thirty-five cents left in my pocket and an absolute determination to return to New York University to earn a Bachelors of Fine Arts degree in the Film Program. Living in New York City and attending the film workshop had been a test of my recovery. From that point forward I no longer considered my schizophrenia a major issue. I continued to take the minor tranquilizer, Vistaril for another couple of years. At that point I learned about the deaths caused by the DBI medicine I was taking to control my insulin output and sugar levels and I decided to try getting off all medicines. I did so with no noticeable changes in my health. To this day I still take a regimen of vitamins, but at much lower doses than I did during those first years. I continue to control my intake of sugars and I stay away from coffee, alcohol and other things which could upset or destabilize my body chemistry. I eat a balanced diet and I make a point of exercising. This burns off excess emotional energy and also helps to keep me grounded in my physical body.

CHAPTER 33
Life in recovery

Once I had gotten though six weeks of the summer production course and big city living without the support of my weekly therapy, I felt empowered to pretty much get on about living a relatively normal life. It would not be accurate to imply that I lived "Happily Ever After" for I faced at least as many of life's challenges and difficulties as most people do. It would also be untrue to claim that I was entirely free from my anxieties or other symptoms from that point on. It is true that I have lived a very full life and that I have handled many difficult problems better than I would have been able to had I not gone through the schizophrenia.

Upon returning from the film course, I stayed for a short time at my parents' house while I found a temporary job at a building supply store. A few weeks later I took a civil service test for the highest paying job I could find in the listings that didn't require a college degree. I achieved the highest test score among the applicants, without using the fifteen civil service preference points to which I was entitled as a disabled veteran. I was told by one of the County Freeholders that among the others I had beaten out were six municipal police chiefs who had all taken the test hoping to retire into the job I won. I would be working with all of them in the new job, but then I wasn't planning to stay forever. I became the County Traffic Safety Coordinator at a salary of $8643 a year and moved into a shared apartment in an old house in the city of Trenton.

I began the process of applying to NYU. I approached the Veterans Administration and opened a new claim to try and increase my disability rating. If I could get my percentage raised from ten to thirty percent I would qualify for vocational rehabilitation and the VA would pay full tuition, disability compensation and a small stipend to attend college. It was the only way I could afford NYU and because I was certain this was the next right step for me I had to try again even though the VA had denied me twice before. By fall of the next year I had been accepted at NYU and with just a few weeks to spare the VA finally raised my disability rating to fifty percent and funded my education.

I handled college well, but there were a few times when some of the extreme and overwhelming feelings that were linked to the schizophrenia would return. Usually they were precipitated by events that were likewise extreme. Unable to resist my dislike of bullies or my Military Police background, I intervened in a mugging on a bus when three youths tried to beat and rob an elderly woman. I reacted as well as anyone could in the moment and protected the woman but fears that were out of proportion surfaced for a time afterward. I was hit in the head with a large rock in the fight with the three muggers and received a concussion which gave me headaches for several months. The doctor warned me that if certain symptoms arose I could be at risk of dying and would have to go to an emergency room immediately. At times, when the headaches came, I could become obsessively concerned about my condition and overly anxious. While my anxiety would sometimes reach high levels I did manage to maintain a sense that I was temporarily out of balance rather than lose my sense of reality as I had during the worst of the breakdown.

Later I saw several people injured and a man killed with a baseball bat during a racial attack in Washington Square Park, by young men from nearby Little Italy. I was followed for many blocks by two black men who I believed were looking to kill a white man in retribution. They confronted me and made threats as I left the park then stayed a consistent distance behind me even as I changed walking speeds several times and led them in a circle to be certain whether they were following me. Finally, I ducked low into the entrance of Brentano's bookstore as I rounded the corner there and slipped down the stairs to the lower level where I would not be visible from the windows to the street. While the paranoia that arose in me as I tried to lose the stalkers was, I believe, well founded, the levels of fear I experienced were nearly uncontrollable and lasted for a couple of days more. Again in spite of the extreme emotions and paranoid thoughts that arose, the saner part of me kept things in perspective to a great extent and I regained control more quickly than would have been possible two years earlier.

I graduated two years after I had transferred to NYU and was granted a BFA in film production with highest possible honors. During my senior year I took eighteen credits a semester and worked thirty-four hours a week as a teaching assistant to the head of the school and as a technician in the film equipment office. By the time I left school I had moved out of the dormitory into an apartment on my own.

Having to face the everyday realities of big city life was helpful for a time as it helped me transition out of the long period when I had been so lost in my thoughts that reality had seemed distant and insubstantial. Nonetheless, it became clear to me that I was more sensitive to input than many people. This may have increased during my breakdown but more likely had been my nature from birth. Either way, the constant stress of city noise and the concern over street crime did take its toll on me. Crime was rampant in New York City at that time. Most everyone I knew at school had been mugged, burglarized or sexually assaulted. We were told that there was an average of one rape a week inside the university library! The perpetrators were often muggers who gained entry using stolen student identification cards, which had no pictures at the time. High crime in the area was the motivation for the racial attacks I witnessed in the park by the citizens of Little Italy who took a vigilante approach to the job they thought the police were failing to perform.

I lived in Greenwich Village on Eleventh St. It was considered a desirable neighborhood. The two blocks of Eleventh on either side of Fifth Avenue had a block association that published a monthly newsletter which included a listing of reported crimes on the street. Most months there were between thirty and forty muggings, car break-ins, burglaries and other crimes in just those two blocks. The building I lived in was broken into on more than one occasion. Thieves came in through a roof access hatch next to my door on the top floor. They broke into the apartment below mine and completely cleaned the place out taking even the man's dishes and linens. A year after graduation, having worked several freelance film jobs in states and cities that seemed safer, saner and less stressful, I moved to California. I packed my film tool kits, my clothing and not much else into a 1966 Plymouth and set out across the country.

Over the next few years I worked as a free lance film technician learning many aspects of the business. Eventually I was given a chance to advance and I became a film and video director. I spent many years directing, writing and producing training and promotional films for Fortune 500 companies, television commercials for regional markets, public service announcements on drug education and sexual assault and lobbying films that were shown to members of congress that have impacted auto safety and medical legislation. Later in my career I became involved in national and international infomercials and worked as a Supervising Producer, Creative Director and Vice President of Operations for two of the biggest marketing companies in that field.

Soon after I first moved to California I found it difficult to stay employed regularly in the film business and for a time I explored a second career possibility. I took a few summer courses in psychology at U.C.L.A., worked as a research assistant for a professor and a graduate student in the psychology department and took a part time position as an assistant teacher to adolescent inpatients at U.C.L.A.'s Neuro-Psychiatric Institute. I took a battery of tests including the standard Graduate Record Exams, the Advanced Psychology GRE and the Miller Analogies Test and scored extremely well. I applied and was accepted into graduate degree programs at the University of Washington and at Vanderbilt University. I moved to Nashville, Tennessee and entered the PhD. program in Clinical Psychology at Vanderbilt where I was one of eight students accepted from a pool of four hundred and eighty plus applicants.

I earned excellent grades during the first semester but I did not complete the program, a decision I struggled with as much as anything in my career track. I was advised by several people I knew in the psychology world to hide the fact that I had been diagnosed as schizophrenic from my professors and the other students. I was told the academics would immediately identify me as someone who had "gone native" and become unable to see me objectively. This may have been the correct advice but it added stress as I tried again to sublimate my real self to fit in. The program was less clinically oriented than I had hoped and I felt that the graduate students were essentially being used as research assistants to the professors, precious few of whom had any significant clinical experience themselves. What little clinical training existed in the program was based in behavioral psychology, an approach largely founded by an ivory tower psychologist whose theories grew out of having his graduate students give electric shocks and food pellets to caged pigeons to change their behaviors. There was virtually no acknowledgement within this approach of what was happening subjectively to people that were disturbed and having been through what I had been through, I found it increasingly difficult to keep silent as I read through dozens and dozens of books and reams and reams of research papers that I felt had virtually no understanding of what actual patients went through or what they needed. I was one of four of the eight in the incoming class who were disappointed enough to drop out at the end of the first year.

I went back to work in film with the intention of possibly finding another program at a later date. I began to have some real success in the film world making commercials, medical and business films and never returned to academics. It is a decision I still question.

In the area of relationships, I had as many joys and disappointments as anyone else. By my senior year at New York University I had developed enough confidence to become more outgoing socially and I dated often. I became involved with a girl in the class behind me that was from Los Angeles and we moved in together in California. It was an eight year relationship that ended in a six month marriage. We had it pretty good for most of the time but in the end there were issues about a couple of things including when and if to have children. By the time we were married I was thirty-eight and wanted children. My wife was thirty-one and wasn't ready.

About that time, I met a younger woman who said she had been pushed into being a singer but instead really wanted to have a family. My first wife divorced me and I was married to my second wife almost immediately. We had a stormy five year marriage which ended in a twenty year divorce; at least it seemed that long. Coincident with the birth of our first child, my second wife decided she wanted to be a singer after all. We had a second child two years later and by the time he was one and a half we were divorced. I worked at staying close to the kids and providing support, love and guidance. The divorce and the challenges of staying close with the children in spite of the strained relationship with my ex-wife are among the biggest tests of my ability to handle stress after the breakdown. I believe I have been a good dad and I am certain I have worked hard at it.

As part of the health regimen I followed through the post breakdown years, I kept a strict diet and stayed away from caffeine, sugars and alcohol. I did not have a single alcoholic drink for fifteen years since the beginning of my psychosis. Around 1985 I tried to drink moderately again but as the stress began to build in the deteriorating second marriage I used alcohol to suppress my feelings and my drinking became a problem. In September of 1992, about a half a year half into the divorce I signed myself into Betty Ford Center, a place for which I had previously made medical and educational films. I never took another drink and I am happily still alcohol free more than twenty-three years later.

The counselors at Betty Ford told me I did not fit the profile of a classic alcoholic but rather that of an Adult Child of Alcoholics. They said I am more sensitive or highly strung emotionally than some people, that I came from a background where I became acutely attuned to other people's needs and moods to avoid being raged at or mistreated. I had learned to make my feelings unimportant. At Betty Ford I was reminded again that if I don't attend to and honor my own feelings, if I hide from them by overworking, or drinking, by trying to please everyone but myself, or by using any number of distractions, my illness could reappear.

In my fifties and sixties, I again faced serious life challenges. On one fall morning around 5:30 am west coast time I was sitting alone on my living room floor in Palm Desert, California. I was doing an exercise I do from time to time that involves reviewing any resentment I might be holding and then shifting focus and visualizing forgiveness and positive energy flowing from me. The idea is to start small with people you care for that are close to you and then to slowly expand feelings of love or positive energy to as wide a group as you feel capable of sending good thoughts to. That particular morning I started by visualizing my children receiving good energy then expanded to include their mother. Soon I could see love spreading to my sister, my mother who lived back east and then to my nephews and friends in New York City. It was a particularly powerful and focused meditation and I felt unusually able to get my self centered thoughts out of the way and visualize good energy going to a much wider than usual group of people, centered around New York where I had lived and gone to school years earlier. I had no particular reason for focusing in on New York, it just came up.

I finished my meditation practice and drove to a local golf course. As I left the pro shop with the intention of going to the first tee, I passed by a TV where I saw the second airplane impact the World Trade Center towers. It was September 11, 2001. I went home and called my good friend from college who lived in lower Manhattan. We talked on the phone as he filmed the horrific event from his apartment window. We later learned that his girlfriend who was eighty some floors up in the first tower when the plane hit, had some kind of intuition to break off from the group she was trying to evacuate with and go down a different stairway. Everyone that went down the first stairway ahead of her was killed and she was in fact the last one to escape on the stairway she did use.

The 911 attacks brought back many strong and difficult feelings about war. When the US went into Afghanistan and later Iraq I struggled once again with my own memories from the Vietnam War. When my eighteen-year-old son went into the military and became an Army Ranger deployed in Afghanistan with a unit assaulting time sensitive targets several nights a week, I entered therapy once again recognizing I needed some additional support. He completed two tours successfully but on his third tour his platoon had many losses. He lost his best friend and mentor who died in his arms. My son was EMT trained and suffered deeply over not being able so save his dearest friend. Soon after that his team lost another member and had more than a dozen wounded.

My son's strike force included his own Ranger platoon and a similar sized contingent of Navy Seals. They tented together, planned operations together and worked together in the field on a daily basis. The Seal unit was the part of Seal Team Six that was shot down in a Chinook helicopter with no survivors. My son's unit had to break off an assault and fight its way through an imminent ambush to secure the crash site. He spent two full days and nights with the charred and commingled bodies of his friends and fellow warriors. He was hospitalized and treated for Post Traumatic Stress Disorder on his return to the states. I visited him in the hospital in Oregon and he seemed to be making progress. After his release I had trouble getting him to stay in touch and I became concerned that he was discouraged and could be resorting to drinking or taking drugs to conceal his pain. I contacted his unit's family liaison person and she assured me his superiors were aware of his struggles and would make a special effort to monitor his progress.

In June 2012 I sold my home of twenty years in Palm Desert, California. I left a call for my son to let him know how to reach me then drove to Los Angeles to say goodbye to my daughter. The next morning I headed east toward South Carolina to begin a new life in retirement with my third wife, a childhood friend I had reconnected with in 2005 and married in 2010. We got as far as Oklahoma two days later, when I received a call from my son's commander telling me that my dear and only son was in a hospital dying from a self-inflicted gunshot wound. A few hours later as I was frantically trying to find a way to fly back across the country I learned my son, Sergeant Morgan Bixby, had died at the age of twenty-two.

To make matters worse, over the next many months my twenty-four-year-old daughter began to hover on the edge of serious disturbance due to her brother's suicide. Life has presented me with many tests over the forty some years since my own breakdown. This has been surely the hardest. I frankly don't know where I would be today without the strength I gained in my recovery from schizophrenia to fall back on and to share with my daughter, wife and other family members when my son killed himself.

CHAPTER 34
Epiphanies

My breakdown began with a descent into a world of terror, darkness and foreboding from which I believed there was no escape. Reality as I had previously understood it ceased to exist. My colorless world was made of gray shadows that barely concealed the unspeakable and ultimate evil which I was afraid was the source of all things. I became numb with fear; numb to the external world, to my body, my emotions and to everything else except the evil blackness that lay in waiting at the edge of all my thoughts.

Today the world is a very real place to me, sometimes excruciatingly real as should be evident from the previous chapter. My journey back to reality was not easy and my life since that time has not always been simple or entirely pleasant but for the most part I have been able to appreciate life in ways that would not have been possible if I had not struggled through the schizophrenia. That is what gave me the tools and the courage to meet the challenges of life head on. I also now have a far better understanding of what is important and what is not.

The underlying causes of my schizophrenia are likely numerous and complex. The traumas of childhood and war and my confusion and shame over sexual feelings are some of the areas where feelings were so strong that I was unable to accept them consciously. My alcohol use at an early age and my brief experiment with drugs assisted me in anaesthetizing those feelings. Strong threats of punishment for not conforming at Catholic school and at home contributed to keeping my true nature well hidden, even from myself.

As a child I felt under constant threat of bullies and blowups by the nuns and my parents and as a result developed a kind hyper vigilance that fed my paranoia when I lost my bearings in reality. The feeling that I was often at risk of being harmed became a part of me and caused me to be more introverted, lost in my thoughts and less engaged in the more threatening external world.

On top of all these factors, my body and brain chemistry were off balance at the time of my breakdown. Drinking, drugs, weight loss from the malaria symptoms, gastroenteritis, and physical exhaustion from more than a year at war were all contributing factors.

I initially got some temporary relief from symptoms when the first two doctors tried to alter my body chemistry with medicines and diet. The limited sense of stability resulting from these efforts was ultimately undercut and I relapsed into deeper terror when the medicines failed to suppress my symptoms over the long term and my condition even worsened due to side effects from those very same drugs.

When Dr. Jane Rittmayer confided her own personal experience with schizophrenia, I first became open to the idea that there could be an end to my suffering short of the annihilation of my mind. The actual healing process first began when I started to remember the combat dreams. The core of that process was extremely simple. I learned to attend fully to my schizophrenic symptoms rather than to deny or dull them as the drug therapies tried to do. My fear and other symptoms were messages from a deeper part of me. I had to begin to listen to my natural or primordial self which had been there all along but had been ignored by my ego or internalized self, a false self which had built up over time to cope with a world that was unlivable to me.

As I risked more and more to acknowledge those feelings and messages from inside, I had more and more relief. The more I committed to this excruciating but very simple process, the more the road to healing seemed to unfold on its own and led me up and out of the hellish world of schizophrenia.

As my fear and the other psychotic symptoms left me, my self awareness increased exponentially. I became comfortable with who I was at a very profound level. I developed the confidence to step out into the world and risk having strong feelings that were both positive and negative without fear of losing control. My emergence from psychosis became the entrance to a lifelong path of personal growth and development that eventually took me beyond anything I might have hoped for as a child or a young adult.

Along the way I had many "aha" moments or epiphanies. Some of these were mundane. Some were powerful. A simple example of such an insight had to do with my feelings of loneliness. On New Year's Eve of 1972 I was still living in the small room in my parents' home. I had almost no friends, little social interaction outside my therapy sessions and certainly no place to go to celebrate the occasion. My parents had a noisy and well lubricated group of twenty or more over to ring in the New Year. I was embarrassed about still living at home and especially about not having a date or even friends to be with. I sat in a spare room upstairs watching the rest of the world celebrate on a small portable TV. There was a steady stream of partiers making their way up and down the stairs to use the bathroom just beyond the room I was hunkered down in. I had tried to close the door but my mother had insisted it be left open and had given me a lesson on heat distribution to make sure I understood the reason for doing things "her way." I ended up watching TV crouched on the bare floor, partly hidden behind an armchair to avoid being seen by the guests. Hiding didn't help as everyone saw the flickering TV in the darkened room and poked their head in to see who was watching it and to ask me essentially why I was such a loser that I had nothing to do but hide in the corner on New Year's Eve.

The "aha" moment came later when I looked back on the uncomfortable feelings of that night and came to the realization that the discomfort came from the embarrassment rather than the actual loneliness. I was far more afraid of being thought of as lonely than I was of being alone. From that point on I no longer resisted the feelings of embarrassment. I found being alone more tolerable and sometimes even pleasant. I became less afraid of failure when I tried to date or make friends and as a result became more available and much more successful socially.

Attending to and investigating feelings that were initially vague or puzzling has been the key to unlocking these long hidden limitations that I had accepted or created. Breakthroughs continue to this day, years after my schizophrenia symptoms disappeared. They often develop spontaneously once I simply put my attention on uncomfortable feelings. While living in Los Angeles I used to go often to Santa Monica and walk along the park at the palisades there. The palm tree lined walkway follows the edge of a hundred fifty-foot-high bluff that overlooks the beach and the Pacific Ocean. It is a beautiful spot and there are usually dozens of local regulars as well as many visitors from around the world relaxing as they stroll along, soaking in the beauty. One day I began to recognize that I was far more tense than most of the other people using the park. I seemed to be more focused on getting through my exercise regimen than on relaxing or enjoying the scenery. I began to look deeper at my state of mind and found the old hypervigilance at the edge of my thoughts. I had been beaten so regularly as a child and attacked so many times when I was not expecting it that I was carrying with me an irrational fear that someone might suddenly come at me and try to throw me over the cliff. This of course was a projection that had no basis and the very moment it came into my consciousness I was able to understand its source and then dispel it. These moments truly brought a euphoric sense of relief and freedom. They made me aware of choices in my life that I had been unable to take advantage of because of the hidden fear, anger, sadness and other feelings that I had restricted myself from feeling.

I found that once the feelings surfaced they were almost always manageable, if not immediately then at least after repeated attempts. The problem which had to be overcome was that at an unconscious level I was still experiencing feelings with the immature and undeveloped sense of self of a child. When I was able to bring those same feelings into the awareness of my adult and more solid sense of self, they lost their power. One afternoon I had been hiking in the foothills over Will Rogers Park, above Sunset Boulevard. As I returned toward the parking lot I stopped to watch people playing and picnicking in a large grassy area that served as a polo field on the weekends. I saw a young boy run to his mother and hug her as she hugged him in return. My eyes filled with tears. As I stayed with my sadness I became aware that I had never known that kind of affection in my childhood. That simple insight almost instantly allowed me to begin experiencing my needs for affection as an adult male instead of as an unfulfilled child. This of course had a great effect on my ability to relate to women.

There is one area of feelings that I have never been able resolve entirely even though I have tried for many years to work through it in therapy, in support groups and through writing about it. Even at this moment as I attempt to write this passage my eyes fill with tears and my body knots up in pain. My right side tenses. My head and neck throb and pain runs around from my right shoulder to my index finger. My rib cage scrunches in, my lower back and my legs tense and my teeth clench together. I am having a physical memory or flashback of being in a kneeling firing position in 1968 as I fire my M-16 rifle again and again into a cluster of three Viet Cong soldiers, my deadly rounds driving them down and down and one of them stares directly into my eyes as his life's blood turns the thick dust of the roadway into a five-foot puddle of deep red mud.

When I try to stay with those feelings, they intensify. I sob uncontrollably. I picture one of the dead as a boy like myself at that age, a boy with an artistic temperament who found himself in harm's way before he had reached adulthood, before he had known who he was or why he was doing what he did. I fear for his soul and I begin to imagine the grief of his parents, his siblings and his friends… and I weep.

I have learned to uncover and penetrate many difficult, confused and even excruciating feelings over the many years since my breakdown. The sense of freedom, relief and transcendence which comes from that work would have been unimaginable to me until before I experienced it. The feelings associated with my first personal kills during the war, however, still to this day have as strong an impact on me as they did when they first surfaced through my nightmares. I have learned to live with them from time to time and I have stopped trying to fix them or change them because I have finally accepted that having to kill people really is that goddamn sad.

The many horrors of war that I witnessed and the cruel and sometimes vicious treatment I received as a child certainly contributed to my doubts about the goodness of the universe or its creator. My obsessive need to investigate the nature of reality grew from that and led me through a dark and terrifying maze of possible explanations, none of which was rationally satisfying or empirically verifiable. I was no more successful in finding final answers to questions about reality or its underlying order than the philosophers, scientists or theologians who have struggled with those same questions throughout recorded history.

What I did find as I journeyed through my schizophrenia was the answer to the question that had seized my attention that chilly evening on the drive to Peddlers Village about who had been driving the car during the gap in my attention. That question, as it turned out, was an extraordinarily important one. It pointed to the now obvious fact that I had no idea who I was. Up until my breakdown I had functioned by temporarily identifying with personality fragments that were situational at best. These fragments or partial personalities had been conditioned by interactions with nuns, parents, teachers, bullies and by military discipline. Many of the behaviors that fit those situational personas were far removed from my natural or original self which I discovered by working through layers of fear, resentments and other feelings that were so powerful I had created massive, even cosmic sized projections to explain them. Those projections became my focus instead of the real issues or experiences which were the true source of my strife. What finally unfolded was less about whether the universe was good or evil or whether the external world was made of mind or matter, but more about whether I could shed ideas I had internalized about who or what I should be and instead accept my self as who I really am. The big surprise was finding out that every time I dropped an old idea of who I was, instead of feeling smaller, I felt larger. Instead of trying to live up to ideas of who I was, I began to just live.

Schizophrenia was for me a rare opportunity to revisit and thoroughly investigate my conditioning. It caused me to question everything in my life from the very existence of the external world to the nature of my soul. It provided me with insight into the workings of my mind and with a gateway to my true self. My own nature was tested to the point where I no longer need to question who I am. My sense of reality was also tested to the point where I no longer doubt that either. When I finally let go entirely of concepts about the world or myself, both emerged clearly as what they are. I found an earthiness or solidness in me and in the world that no longer needs to be questioned. I found a floor or a sense of ground on which to rest and at the same time a sense of open space in which to live and explore.

I would never wish even the smallest bit of my schizophrenia on anyone else and I most certainly hope with all my heart that I never go through any of it again. I am eternally grateful that I survived it and for the lessons I learned along the way. Schizophrenia has made most other problems small by comparison. It has made each moment of my life in recovery a most precious gift. It has given me compassion for others and the strength to reach out more often with a kind word or a helping hand. It has helped me to discover a measure of inner peace and even to love and accept myself as I am. Schizophrenia has made me an authentic human being.

EPILOGUE
Final exams

My writing and editing work on Crazy Me was interrupted several times by significant and often stressful life experiences. Each of these events tested my stability in recovery. Together they have given me an even deeper appreciation for the strength and determination that my desperate journey through the darkness instilled in me. I can see more clearly how the strategies and techniques that helped me to simply survive the most despairing days of my breakdown, have stayed with me and become a necessary and valuable part of my daily coping skills.

I was nearing completion of Crazy Me when I sold my home of twenty years, packed my belongings onto a moving van and set out across the country to begin a new life in retirement with my new bride, a childhood friend I had recently reconnected with. Two very long days into a five-day drive as we exhaustedly checked into a hotel in central Oklahoma, I got the devastating call from my son's military commander.

"You are aware that your son has been under treatment for Post Traumatic Stress Disorder," he began.

Of course I was. I had flown up to visit him when he was hospitalized in a "Freedom Care" facility in Oregon.

"He has been hospitalized with a self-inflicted gunshot wound."

I had expected him to tell me that Morgan had gone back into treatment for his PTSD.

"Where…" I hesitated, "did he shoot himself?" My mind reeled. If he was in the hospital, he was alive. Maybe he shot himself accidentally, maybe in the leg or foot. I prayed.

"In the head," came the quick and horrifying reply. The major went on to tell me that although my dear and only son was breathing own his own, there was no brain activity.

I turned and looked in through the window of the motel restaurant at my new wife waiting patiently for me and for the fried chicken dinner that had been ordered but that neither one of us would be able to eat.

During the worst of my psychosis I had been haunted incessantly by the overwhelming fear that the world around me would dissolve along with my body and my naked soul would plummet eternally through a black pit of hell flames and sadistic demons. Now my body convulsed as a cyclone of fear, anger, grief and emotions beyond description erupted through me. I wanted the world to dissolve. I wanted to be dead.

Images of my son in his best and worst moments flooded my brain. At the same time, memories of the most excruciating and terrifying episodes of my schizophrenia shattered my sense of time and space. There was nothing to hang onto, nowhere to run or hide. Whatever I had imagined as the future was gone in an instant. There was no escape and I was defenseless against the horror that had just become my reality.

As I started to share the terrible news with my wife, I began to feel a hint of the strength I had gained through my battle with insanity. I knew I had already been lost as deep into darkness as most anyone had ever returned from. I had handled that. I could handle this. I would handle this. I had to. My daughter would need me. My wife, my sister and my mother would suffer as well and I would be the one they would look to.

We made it back to the motel room. I turned out the lights. Somehow it helped. I dialed my daughter. I didn't want her to have to hear the news from anyone else. Having to tell her that her brother was dying was as difficult as anything I have ever had to do.

I called the hospital my son was in and they stonewalled me because I didn't have a "password!" "I'm his father," I insisted. Finally, they put me through to the ICU nurse. "He has no brain activity," she informed me when I told her we were frantically trying to find a flight to be at my son's side. I asked her to place the phone by my son's ear and leave the room. I had seen people die and was firmly convinced that the soul did not always leave at the moment of physical death. If my son was still breathing, I was certain that he was still attached to the physical plane in some way. If there was a remote chance he could still feel my love or if I could impart even a hint of positive energy to help him in his transition, I was determined to try.

I was awake and crying through most of the night. At 5 o'clock a.m. we got back on the road not knowing whether to keep driving toward our destination three days away or to turn south toward Dallas where we would have the best chance of getting a flight to Washington state. My wife was on the phone with the airline when a call waiting came in with news that my son, Sergeant Morgan Bixby, had died an hour earlier.

The next three weeks were something of a blur. We made it the remaining twelve hundred miles across the country, sometimes crying, occasionally screaming in agonizing grief. There were moments when I thought I might spin the steering wheel and accelerate the Corvette off the road and into oblivion. I kept telling myself that I had survived schizophrenia and I could survive this and I kept telling myself I had to stay on the planet and keep it together for my daughter, my wife and for the rest of my family. Our house was not ready for us when we arrived in South Carolina and the moving van with all of our clothes had broken down, multiple times it turned out, so we ended up buying clothing for the funeral at a thrift store.

I initially did not think I would be stable enough to fly back across the country in time for the services but with the help of the Army's Survivor Outreach Services and in spite of several glitches including a cancelled flight that left us stranded in the middle of the country; I honestly don't remember where, we arrived in San Francisco and checked into the same hotel where we had stayed and put on the rehearsal dinner for my son's wedding less than six months before.

As the day of the funeral drew nearer deep feelings of fear, anger and hopelessness battered me from all sides. I began to worry about what would become of my son's consciousness, his soul… all of our souls, in the afterlife. I was angry at the people who had had responsibility for Morgan's medical care and at the people that had put him through the horrors of war that had caused his problems. At moments I became despondent that I had never… would never get anything in my life right. I had had two failed marriages, a mediocre film career and had not finished graduate school. The one thing in my life I believed I had done right was raising my children and now one of them was dead from suicide and the other devastated and hovering near the edge of serious disturbance herself as a result.

Thankfully, the strength and the techniques I had learned from my journey into and back from insanity were with me. I knew to stay with the feelings no matter how intense. I knew not to follow the convoluted thoughts my mind projected to distract me. Imagining how I could have saved my son wouldn't bring him back. Projecting how the psychiatrist and others who failed him should be exposed or punished would help no one. Obsessing for hours, days or months about the horrible thoughts the Catholic nuns had implanted in me about suicide would have no impact on Morgan's actual experience in the afterlife.

I made time to be alone with my feelings, to explore them and to separate myself from the web of thoughts that was trying to cover them up. I knew community was important and I reached out to the Vietnam Veteran brotherhood through the Vietnam Veterans of America organization. The California VVA President, Steve Mackey, and his wife, Elayne, drove a thousand miles round trip to stand beside me as my son's remains were unloaded from the plane at the San Francisco Airport. They helped arrange for three VVA Chaplains and two hundred flag carrying motorcyclists from the Patriot Guard to escort us and to stay with us through the burial.

The single most important lesson learned during my breakdown and recovery probably had the most to do with keeping me on track during this period as it had done many times before. There were many other people suffering, my daughter, wife and sister; my son's wife, his mother and so very many of his friends and fellow soldiers. As I shifted my attention to them I found not only that I could be helpful in some small ways, but that my own energy returned and with it a determination to go forward and try to make the best of what we still had. Recognizing that how I handled myself would affect my family and friends helped me to stay positive.

I had journaled and made notes on the trip back across the country to keep me from melting down when the plane was cancelled and during the other interruptions of our travel. I continued to make notes to myself to help me stay in contact with my feelings. Those notes ended up as the basis for a fifteen minute talk I somehow found the strength to give before a group of four hundred mourners that gathered at an American Legion hall after the funeral to honor my son's memory. The message I delivered was one of gratitude to all the people who had tried to help my son over the years. I told them that Morgan had been on a path of his own since the moment of his birth and no one should feel that they could have done anything differently that would have changed the outcome. It was a message that I understood and believed intellectually but that would take me much longer to emotionally accept, regarding my own responsibilities to my son.

We returned to South Carolina and lived in limbo for another two weeks before our new home was ready. The moving van took yet another week due to the breakdowns. We were not in the house one week when I began having severe anxiety attacks. We had unknowingly moved into a neighborhood where I could hear the gunfire from a nearby military training base! I already had what is known as an exaggerated startle response due to my PTSD issues and now the sharp report of each weapon reminded me that my son had taken a gun to his head. We put the house on the market.

About three weeks after my son was buried my wife's mother became terminally ill. We moved in with her four to five days a week and helped with her care during her final four months. Once again, the opportunity to help someone else helped to shift my attention away from my own loss and do something that helped restore my sense of worth. The year or so that followed was filled with more loss than either my wife or I could have imagined. Shortly after my mother-in- law's death my best friend from high school, who I had visited with days before, died from a sudden and fatal stroke. Soon, my golfing friend of many years died from emphysema, we lost a niece to cancer, my own mother died, two dear friends contracted long term and debilitating illnesses and my best friend from college committed suicide after a long battle with kidney disease. I continued to be concerned over my daughter's emotional state and as if providence was determined to test the outer limits of my stability in recovery, I also went through two eye surgeries and suffered a back injury that took four months to heal.

For more than a year since my son's death I hadn't been able to write a single word in spite of the fact that *Crazy Me* was nearly complete. Maybe I felt that if I couldn't even save my own son, the thing I wanted more than anything in my life, that perhaps I had nothing of value to say. I couldn't save him and now after the year we had just gone through, I was clear that I couldn't save anyone else, but the image of the young psychologist, Dr. Jane Rittmayer, kept coming back to me. She had quietly shared her own struggles with schizophrenia with me and I remembered it was at that very moment that I first felt hope that there was a chance I could regain control of my own life. I thought back to the many times over the years when the memory of that simple and pure moment in her office had inspired me to keep going. "If she can do it, maybe, just maybe I can too."

And suddenly I threw myself back into finishing the book with a passion. I understood now that it wasn't up to me to save anybody. Now instead, I hoped that if I could share what had happened to me, if I could tell about what had worked for me... maybe I could give hope to someone somewhere... that they could help to save themselves.

ACKNOWLEDGEMENTS

Thank you to Jane F. Rittmayer, EdD for being open about her own struggles with schizophrenia. This was the single most important moment in my recovery.

FURTHER READINGS

The following selections cover material outside popular current medical models of mental illness but were helpful to me in understanding my process. Many are out of print but can still be obtained used, in electronic form or in libraries. My comments follow each selection.

How to Live with Schizophrenia – Citadel Press, 1979 by Abram Hoffer, M.D., Ph.D. and Humphrey Osmond, M.R.C.S., D.P.M. — A comprehensive biochemical theory of causes and treatment of schizophrenia. Most currently practicing psychiatrists I have spoken with are unfamiliar with this theory, but its early adherents were most helpful in stabilizing my condition in the beginning stages of healing.

The Schizophrenias: Yours & Mine – Penguin Group (USA), 1980 by Carl C. Pfeiffer Ph.D. M.D., Jack Ward M.D., Moneim El-Meligi Ph.D., Allen Cote M.D. — Co-written by the psychiatrist who helped me through the first stages of healing.

Schizophrenia as a Human Process – Norton, 1962 by Harry Stack Sullivan, M.D. — Dr. Sullivan was among the first to think of mental illness as a process that could be navigated with a positive outcome.

Notes to Myself – Bantam-1990, by Hugh Prather. — This somewhat poetic book of internal dialogue helped me to realize I had issues much closer to home than my concerns about eternity and my place in it.

On Becoming a Person – Houghton Mifflin, 1961 by Carl Rogers, PhD. A seminal work on "Client Centered Therapy." — My therapist, Dr. Rittmayer, adhered in large measure to this approach with great effect in my own case.

Memories, Dreams, Reflections – Chapter VI "Confrontation with the Unconscious," Vintage Books, 1965 by C.G. Jung. — Jung's encounter with "cosmic stillness" and his insights into the unconscious through his own dreams and artwork parallel my own experiences.

Adult Children of Alcoholics – Health Communications, 1990 by Janet G. Woititz, EdD. — Insights from this book about personality tendencies of those raised in alcoholic families continue to help keep my growth on track to this day.

Psycho-Dietetics – Bantam, 1975 by Dr, E Cheraskin and Dr. W.M. Ringsdorf, Jr. — An early work offering dietary answers to emotional issues.

Character Analysis – NoonDay, 1965 by Wihelm Reich, M.D. — Largely discredited by his medical peers for his belief in invisible Orgone energy, Reich offers unique insight into a process he calls "armoring," which well describes the extreme physical tension in my own body that walled off unacceptable feelings.

ABOUT THE AUTHOR

As a survivor of PTSD and Schizophrenia, D. Thomas Bixby earned a B.F.A. at NYU, and studied psychology at UCLA and Vanderbilt. He is an award-winning writer and director of over a thousand commercials and medical films and has done volunteer work with veterans, recovering alcoholics and the mentally ill. Club Bong Son, his script about PTSD won Best Screenplay at the 2015 Beaufort International Film Festival. He belongs to the DGA, Mensa, Triple Nine Society, Vietnam Veterans of America and is a life member of the Disabled American Veterans.